ALWAYS IN THE VANGUARD: THE BUFFALO
SOLDIERS OF COMPANY H, TENTH U.S.
CAVALRY

ALWAYS IN THE VANGUARD

*THE BUFFALO SOLDIERS OF COMPANY H,
TENTH U.S. CAVALRY*

BY JAMES T. MATTHEWS

Starfire Press
San Antonio, Texas

Library of Congress Control Number: 2023914677

ISBN 979-8-218-28362-9

Cover art is from a drawing by Bonnie Curnock in 1996. The photograph on the back cover is from *Under Fire with the Tenth U.S. Cavalry* by Herschel V. Cashin et al., 1899, page 87. Unless otherwise noted, all additional photographs are either from the public domain or from the author's collection.

Starfire Press is an independent publisher based in San Antonio, Texas. This book was printed by Brenner Printing & Mailing, 1234 Triplett, San Antonio, Texas 78216.

TABLE OF CONTENTS

INTRODUCTION

It was at a symposium on the "African-American Military Experience" in the spring of 1998 in Carlsbad, New Mexico, that I first thought of putting together some type of comprehensive work on Company H of the Tenth Cavalry. I had worked on the committee organizing that symposium and prepared a paper concerning Buffalo Soldier patrols and campaigns in the Guadalupe mountains. In doing so, I had a chance to look further into the exploits of Company H, later known as H Troop of the Tenth U.S. Cavalry. I realized that this company had been involved in every Indian campaign on the Southern Plains and had remained an integral part of army operations up to World War I. Set against the overall history of post-Civil War American expansion, I believed that Company H had a very important story to tell concerning African-American military service in the first generations following freedom. It took many years of research and a few setbacks in publishing before I completed this project. Through it all, my wife Becky continued to provide inspiration and encouragement to keep me going. It is to her that this book is dedicated.

In the years following the American Civil War, it must have been a glorious thing to be free from being forced to work for one person in one place no matter what your desires, to be lifted up from being debased as property or something less than human, but freedom did not give the former slaves food to eat, a decent job or the respect of those who had been accustomed to humiliating them. Those were things that they had to find for themselves. It did not prove to be easy, because in the minds of many people, they did not start out on an equal footing with other citizens.

Yet the United States Army did provide one road to opportunity. The army had already given many immigrants and sons of poverty the chance to make life better despite their circumstances. In the mid-1860s, it provided that same chance to young Black men, mostly lower wage laborers in their twenties. Of course, the army still had its share of prejudice in matters such as race, class and national origin. But the daily routine of a frontier outpost did provide a better atmosphere for growth than many communities of that time, especially in the post reconstruction South. On duty, the soldier's greatest challenge was to survive the long patrols, the endless hours of guard duty and the mediocre rations.

They were assisted in that endeavor by a few good officers. While all of the white officers had some prejudices, Company H was fortunate to have many officers who genuinely cared about their men and tried to make them the best soldiers possible. Captain Louis Carpenter, one of the best company grade commanders of the period, stands at the top of that list. Although he sometimes despaired the Black soldiers' lack of education, he put forth an extra effort for his men and treated them just as he would have treated any good soldiers. When I read about Carpenter, I was reminded of my own experiences. A few years back, I was assigned as the Boy Scout Executive for a district area that was 80% Black. There were occasions when I was the only white face present, such as the time I gave a homily for Scout Sunday at an AME church or when I kicked off the Fourth of July program at a community center. But I always felt my job was worthwhile and in time the people of that community became important to me. I believe that is how it must have been for many of the officers of Company H.

The soldiers of Company H came from a wide variety of backgrounds and so provide an excellent portrait of the African-American soldier during the last half of the nineteenth century. Many of those who enlisted in 1867 had fought during the closing years of the Civil War, most of them in the United States Colored regiments. All of them had been born in the United States, unlike many soldiers in the white regiments, who were recent immigrants from Europe. Although many Company H soldiers had been born in former slave states, almost a quarter of the recruits came from Pennsylvania and the Northeast. In the later years of the nineteenth century, more Black soldiers were recruited from the Midwest and other regions. Many could not read or write and so had only been able to hold menial jobs. In this, army service proved to be of benefit. Officers encouraged basic education for all soldiers on post, largely because they wanted their senior enlisted men to handle the large volume of forms and documentation required by the army command.

Looking at the history of one company allows for a closer study of the culture and attitudes of the Buffalo Soldiers to compare them to those of the army as a whole. Their service can be observed against the background of a constantly changing society as the United States struggled into the twentieth century. Although victims of constant prejudice from the very beginning of their service, conditions did become progressively worse after the Jim Crow laws were enacted. Troops that had served with great distinction were shunned by local citizens and the army became reluctant to use the Black regiments in certain situations. Economics also played an ongoing part in recruiting and retaining soldiers, with the greatest increase in enlistments always occurring each time the national economy began to weaken.

The story of Company H links the service of Buffalo Soldiers from the historical past to our present day. Reuben Waller, who joined when the company formed in 1867, died on August 20, 1945, just two weeks after the first atomic bomb was dropped on Hiroshima. Medal of Honor winner Dennis Bell died in the same year I was born. The troopers of Company H were a cross section of fairly ordinary men, from thirty-year career soldiers and battlefield heroes to repeat deserters and thieves. In all they built an extraordinary record of service to the U.S. Army in often harsh and difficult circumstances. Theirs is a story worth telling and remembering.

TOWARD A NEW FRONTIER:
THE FORMATION OF COMPANY H

Self Portrait of Louis Henry Carpenter. This drawing is reproduced with permission of the owner, David P. Harrington. Any additional reproduction is prohibited without explicit permission of the owner.

As the battered soldiers of Pickett's and Pettigrew's brigades retreated across the open fields of Gettysburg on the afternoon of July 3, 1863, eight miles to the southwest, at the village of Fairfield, Pennsylvania, a young Union cavalry officer bravely tried to rally a remnant of his regiment against a brigade of Confederate cavalry. His name was Louis Henry Carpenter, Second Lieutenant of the Sixth U.S. Cavalry, and his unit had been sent behind enemy lines to capture a Confederate supply train reported to be in the vicinity of Fairfield. Already divided into separate squadrons when the Virginia cavalry attacked, the Sixth proved unable to collect their forces and were driven back through the town of Fairfield. Of the Sixth's officers, only Carpenter and Lieutenant Nicholas Nolan remained with the command. The others had been wounded or captured. Three times within half a mile, Carpenter managed to rally over 100 of the troopers. Standing firm in the midst of confusion, Carpenter lost both saber and bridle. The bridle, he soon replaced with one from a captured Confederate horse, yet as he continued to face the oncoming waves of enemy horsemen, he feared none of his regiment would escape death or imprisonment. Carpenter later wrote, "At the fight near Fairfield, I thought several times that I was destined to be an inmate of Libby Prison. But I made up my mind that I would be badly wounded first at any rate. At one time I was entirely surrounded in the town itself, but I managed to cut my way out." With night approaching, the Confederates finally gave up pursuit at the entrance to Fairfield Gap.[1]

Out of the 400 troopers who had ridden behind Confederate lines on the Third of July, 242 had been killed, wounded or captured. Carpenter received the brevet rank of First Lieutenant for his courage in action. Already his reputation as a brave and reliable officer had been well established. Born in Glassboro, New Jersey, February 11, 1839, Louis Henry Carpenter moved with his family to Philadelphia in 1845. He graduated from Philadelphia High School and entered the University of Pennsylvania as a student of medicine in 1859. When the newly formed Sixth U.S. Cavalry began recruiting in Pennsylvania, Carpenter enlisted November 1, 1862. He later recalled, "At the breaking out of the war in 1861, I was engaged in the study of medicine, but being influenced by the spirit which moved the loyal youth of the north at that time, I threw aside my books and entered the army..." He quickly rose through the ranks to sergeant. On September 20, 1862, while suffering from typhoid fever, Sergeant Carpenter received his commission as lieutenant of cavalry.[2]

By the time of the Gettysburg campaign, Carpenter had been in action at Antietam, Fredericksburg, and Chancellorsville. He had also participated in the largest cavalry engagement of the war at Brandy Station on June 9, 1863. During the battle, Carpenter's squadron charged the enemy twice, once to support the Sixth Pennsylvania Cavalry and the second time to resist an attack on their flank. He recalled, "As we went along at headlong speed, cheering and shouting it seemed to me, that the air was perfectly filled with bullets and pieces of shell, shells burst over us, under us, and alongside." Carpenter later wrote to his father, "If I had had command of the squadron, I would have dismounted the men and fought the enemy equally." The Sixth was one of the last units to withdraw from the field and formed part of the rear guard. After Gettysburg, Carpenter received appointment as an aide to Major

General Philip Sheridan and served in that capacity during the Wilderness campaign and the siege of Petersburg.[3]

Sketch of Saint James Church near Brandy Station, Virginia by L.H. Carpenter. This drawing is reproduced with permission of the owner, David P. Harrington. Any additional reproduction is prohibited without explicit permission of the owner.

Now a seasoned officer, Carpenter came to the attention of the anti-slavery forces who advocated the use of Black soldiers in the Union army. On January 29, 1863, Benjamin Rush Plumley of Pennsylvania, a staff officer who had been active with William Lloyd Garrison in the anti-slavery cause, had written a letter to President Lincoln recommending Lieutenant Carpenter for the command of a Black regiment stating, "Too much cannot be said of his courage, sagacity and humanity. You can send no man, who will more efficiently and wisely reach the slave with your proclamation nor more surely enlist him in our service." In October 1864, Carpenter received an appointment as Lieutenant Colonel of the newly designated Fifth U.S. Colored Cavalry. "I organized the regiment, drilled and disciplined it to a good degree of efficiency," wrote the young commander. In all, over 1000 men enlisted for three years' service. Most of them were former slaves recruited in the area around Louisville, Kentucky. Despite his overall satisfaction with the inexperienced troopers, Carpenter did despair of finding qualified non-commissioned officers. To initiate the effective organization of the regiment, he asked and received permission to use some white sergeants that would work with the Black sergeants to maintain order and keep records since "[s]carcely any of the Colored men enlisted in this regiment can read or write."[4]

Carpenter's new command mustered at Camp Nelson, Kentucky, where they patrolled the railroad between Lexington and Covington, guarding against Confederate guerillas. Shortly before Carpenter joined the regiment some of the new recruits had been used in a raid on the Confederate salt works at Saltville, Virginia. This first battle for the Fifth USC Cavalry proved disastrous. On a cold October 2, with their supplies running low, they attacked a well-entrenched enemy. When forced to retreat, some of the wounded left behind were murdered by Confederates who, recalling past fears of slave revolts, had been horrified by the thought of armed Blacks attacking their town. In December 1864, the Fifth, now well trained, organized and equipped, returned to Saltville as part of a successful raid under General George Stoneman.[5]

While Colonel Carpenter prepared his regiment for battle, several hundred miles south in Mississippi, a young Black man named Reuben Waller watched in dismay as United States Colored troops captured and then proceeded to kill soldiers from his master's regiment while shouting "Remember Fort Pillow" and waving banners with that same slogan imprinted upon them. Waller had been with the Confederates attacking Fort Pillow in April 1864 and remembered the horror he had felt then of what he called a "shameful massacre." For three years he had ridden with the Confederate cavalry of the Fourth Alabama regiment as the body servant for his master, William C. Couch.[6] After the engagement in Mississippi, Waller recalled, "I told my master that the black men done our men like we did them at Fort Pillow, and I asked him why it was that way, as they had slaughtered his son and several boys who were raised with me...I could not understand it and master did not enlighten me on the subject..." Despite his shock and confusion, Waller continued to ride with General Nathan Bedford Forrest's southern cavalry, following the only type of life he had known, eventually serving the Confederacy in twenty-nine battles by the end of the war.[7]

Reuben Waller had been born a slave on January 5, 1840 in Wolfe County, Kentucky. Later in life he remembered "the first sensation of my life was the falling stars in 1849. All the slaves and their masters got together and began a mighty fixing for the Judgment Day. Of course our masters had a greater cause for fixing up things, than we poor slaves did, but we followed their advice." The family that owned Waller moved west to Missouri, probably settling on a farm in Cooper County. There the young slave learned to make whiskey, working in the family distillery charring oak bark and making barrels. He later recalled that one of the things that bothered him most as a slave was "white men riding up in the night to Negro cabins, ejecting the men - who maybe later were rewarded with a silver half dollar - and riding away leaving sobbing Negro women." Reuben also remembered, "Well, the next was the '56 Border Ruffian War and the comet in 1860 just before the Civil War. Well, we all got scared at the comet as its tail reached from West to East. It did look frightful. As we think of it now we believed it was a token of the great Civil war and the completing of our freedom, 400,000 of us." Waller rejoiced at receiving his freedom, yet he continued to think of his master's family as his own recalling that the son slaughtered by Union soldiers "was raised at my mother's breast with me..." When the war ended, seeking a life of freedom for the first time, he remembered that he had "engendered a great liking for the cavalry soldiers" through his war time experience and he determined to continue a military career.[8]

Waller was not alone. Many Black men who had served in the Union and Confederate armies saw the peacetime army as a possible career. Many others, both former slaves and free men, saw the military as an opportunity to achieve some measure of equality in a nation still torn by the prejudices brought to the surface through the realities of reconstruction. During the war about 200,000 Black soldiers had served in over 160 units. At the end of the conflict, 65,000 of the U.S. Colored troops were retained on active duty through the spring of 1866 as occupation forces in the southern states. Then as reduction bills began to take their toll on the peacetime army, these regiments gradually mustered out. Colonel Carpenter's Fifth USC Cavalry spent their time following the Confederate surrender along the border of Arkansas and Indian Territory chasing outlaws and unreconstructed rebels. On March

20, 1866, they mustered out of service at Helena, Arkansas. Carpenter took three months leave and reported back to his old unit, the Sixth Cavalry on June 28.[9]

One month later, on July 28, 1866 a congressional act authorized the regular army to recruit six regiments, four infantry and two cavalry, made up of Black enlisted men under the supervision of white officers. Even though many of the troops used to enforce martial law under reconstruction were being withdrawn, congress decided that the use of Black regiments had proven an effective experiment that should be continued. An article in the *Army Navy Journal* during this period commented, "I very reluctantly believe that gentlemen who have made military art and science their study for a lifetime, will let prejudice get the better of common sense." The writer continued, "The colored soldier has been tried and not found wanting. He has gallantly stood up under all the disadvantages of a proscribed race."[10]

Congress designated the two new cavalry regiments as the Ninth and Tenth. To command the Tenth Cavalry, the army assigned Colonel Benjamin H. Grierson, a former music teacher who had led several successful Union cavalry raids through Mississippi during the war. Colonel Grierson reported to the Department of the Missouri in Saint Louis September 10, 1866, and was directed to Fort Leavenworth, Kansas where he would organize his new unit. To recruit his cavalrymen, Grierson looked for experienced officers, giving preference to those who had served with Black troops during the war. The congressional act authorizing two regiments of Black cavalry required that the commissioned officers "shall have passed a satisfactory examination before a board to be composed of officers of that arm of the service in which the applicant is to serve, to be convened under the direction of the Secretary of War, which shall inquire into the services rendered during the War, capacity and qualifications of the applicant." Lieutenant Carpenter, serving with the Sixth Cavalry as quartermaster, had already requested promotion to an open position in a white regiment. Although he probably believed that a white unit would be his best opportunity for advancement, he also applied to meet the board of officers for a position in a Black regiment and readily accepted a commission as captain in the Tenth Cavalry, reporting on December 5, 1866.[11]

When Carpenter arrived at Fort Leavenworth, only sixty-four men had been enlisted in the new regiment. Colonel Grierson, dissatisfied with the quality of soldier he received from the recruiting depots, had established regimental recruiting and training programs using his company grade officers. In this manner the recruits could see the officers they would serve under while the officers would have an opportunity to recruit the most qualified men for their companies. In January 1867, Carpenter departed Leavenworth for recruiting duty in Louisville, Kentucky where a few years before, he had first recruited Black soldiers to organize the Fifth USC Cavalry. Grierson had limited his companies to eighty-four men and called for more selective recruiting. He asked Carpenter to use his experience to "recruit men sufficiently educated to fill the positions of Non-Commissioned Officers, Clerks and Mechanics in the regiment." In March, Grierson transferred Carpenter to his home town of Philadelphia to "enlist all superior men" he could find to fill his company's muster roll.[12]

Grierson's concern about recruiting Black men with usable skills who also were suited to military life proved to be well founded. On March 6, he wrote to Captain H. T. Davis at the Memphis recruiting station, "You will have to foot the bill for your rejects in the future." Despite that admonition, of the twenty-eight men successfully recruited for Captain Carpenter's company at Memphis, ten received disability discharges and seven deserted within the first few years. One, Jerry Williams, was arrested for murder even before he could join the company. Another, Henry Harper, who had been in Carpenter's Fifth Cavalry, died of an unknown illness in October. Even Captain Carpenter experienced problems with some of the men he recruited that seemed to be well qualified. His choice for First Sergeant, John Homager, an experienced soldier at age thirty, became seriously ill in September 1867 and received a disability discharge in December. The oldest Company H recruit, thirty-five-year-old Henry Carpenter of Philadelphia also received a disability discharge in July 1868.[13]

Even so, the time was right to recruit experienced young men with a willingness to learn. Over a third of the recruits simply stated their former occupation as laborer, while almost as many listed farmer. Most of the farmers came from southern states where they probably worked as sharecroppers. Many Black men saw the military as a real opportunity for something better than subsistence living. Robert B. Banks, a waiter from Halifax, Pennsylvania, enlisted with Carpenter on June 20, 1867. He would serve with the regiment for twenty years. Jacob Young enlisted in Louisville, Kentucky on May 24. He had fought in the 24th USC Infantry during the war and would later serve as First Sergeant of Carpenter's company. Pollard Cole, born 1842 in Georgetown, Kentucky, enlisted on June 14 at Louisville, citing prior service in the 12th USC Heavy Artillery, Company K. Cole continued with Carpenter's company as private, farrier and later sergeant, serving the regiment for thirty years.[14]

Almost a third of the men who joined Carpenter appear to have had prior service during the Civil War, most of them with the U.S. Colored Troops. The Black regiments raised during the war had disbanded in 1866 with the end of occupation in many of the southern states. This potentially left thousands of Black soldiers with the freedom to choose their future, but no job. Those who enlisted in Carpenter's company included Trumpeter Silas Jones who had served in the Sixth USC Cavalry. James H. Thomas of Virginia had begun his career as a musician in the First USC Infantry. James H. Clayton, a former corporal in the 84th USC Infantry, came from Richmond where he was working as a shoemaker to enlist at the recruiting station in Washington, DC. Former infantrymen John Clark, Jacob Ewing, Ezariah S. Freeman, Richard Garrison, Thomas Haydon, Alfred McPherson, John D. Price, Charles Sampson and Sidney Sanders, cavalrymen Alexander Adams, Daniel Grissom and Henry Harper and artilleryman Mitchell Jones all enlisted in Captain Carpenter's troop during the spring and summer of 1867.[15]

George Garnett, born a slave in Missouri in 1847, had enlisted in Company I, 56th USC Infantry September 20, 1863, at the age of sixteen. He joined the Tenth Cavalry at Galesburg, Illinois and later became company first sergeant. George Goldsby was born in Selma, Alabama to a mulatto and her former owner. Goldsby could pass as a white man and did so on many occasions. During the Civil War, he had served the Confederate army as a teamster until the battle of Gettysburg. In the battle, he escaped to the Union side and joined the all-white 21st Pennsylvania Cavalry, where he was promoted to corporal. In 1867, he chose to enlist in

16

the Black Tenth Cavalry, although he did not join Carpenter's company until his second enlistment. Henry Allen, a mulatto from Richmond, Virginia had enlisted in Company F, 20th USC Infantry at the age of sixteen. After he mustered out in New Orleans in October 1865, he tried farming for a time and then joined the Tenth Cavalry. Then there was Charles Black, born at Richmond in 1841. In September 1863 he joined the renowned 54th Massachusetts Colored Volunteers while they were attempting to replenish their ranks after the assault on Fort Wagner. He was tried for sleeping on guard duty in September 1864 and sentenced to confinement at Fort Marion, Florida for one year, seven months and fourteen days. Despite his record, Black too answered the call to join Carpenter's company on June 11, 1867.[16]

Many others with no prior service looked upon enlistment as a doorway to a new life. John H. Claggett, age 22, and Joseph C. Claggett, age 21, came to Washington, DC from nearby Prince George County, Maryland to enlist July 9, 1867. Tailor Frank Rogers, engineer John Allen, George Bumpferts and Samuel Jackson all enlisted at Memphis, Tennessee. Carpenter Amos Cormack, blacksmith Daniel Williams and barber Henry Carpenter enlisted in Philadelphia. John Billings, whose experience included kitchen service soon became one of the company cooks. Reuben Waller, now freed to make his own choices, walked cross country with his uncle Martin Reynolds, crossing the Missouri River at Fort Leavenworth. Reynolds settled in the adjoining town of Leavenworth, but on July 16, 1867, Waller walked over to the fort and declared he had come to enlist in the Tenth Cavalry "for the Indian war that was then raging in Kansas and Colorado."[17]

Five days later, on July 21, General Order 15 officially designated Carpenter's troop as Company H, Tenth Cavalry with a full roster of eighty-four enlisted men. Most of the recruits were still in their early twenties. In fact, well over half of the eighty-four men were twenty-one or twenty-two. Only two were over thirty and two were under twenty. Young men trying to find a future in the post war society found a starting point with Company H. The newly recruited cavalrymen seemed well suited for mounted service averaging 5'6" with none of them standing over six feet tall. The majority of recruits came from border states, with over half from Tennessee, Virginia, Kentucky and Missouri. Some enlisted from the Southern states, especially Georgia and Alabama. Due to Carpenter's efforts, ten came from Pennsylvania and there were even recruits from Massachusetts, New York and Maine. Every member of the company had been born in the United States. This proved to be unusual in the post-Civil War cavalry. In the Seventh Cavalry, which organized in Kansas during the same period as the Tenth, over a third of the troopers were foreign immigrants, mostly from Ireland and the German states. Immigrants also appeared to consider the army a good opportunity to make a life in the new country with some regiments enlisting as high as fifty percent of their companies from men who had been born outside the United States. The average age at enlistment in other regiments also seems to have been higher with at least ten percent between the ages of thirty and forty.[18]

By August, Company H had already begun drilling seven days a week, learning the use of weapons and horsemanship. Additional orders named them the "black horse troop." Each trooper should have been equipped with a fresh black steed complete with saddle and tack. Although Colonel Grierson requested fresh mounts and carefully inspected every horse made available to him, open discrimination against the Black regiments surfaced in the matter of mounts and supply. In April 1867, Grierson traveled to Saint Louis to inspect a herd of fifty horses and returned without finding a single one suitable for cavalry service. As Captain Carpenter later complained to the colonel, "Since our first mount in 1867 this regiment has received nothing but broken-down horses and repaired equipment as I am willing to testify to as far as my knowledge goes." He also protested that some of the horses they did receive apparently were rejected mounts from the Seventh Cavalry.[19]

Colonel William Hoffman, the post commander at Fort Leavenworth, had served as an army officer since 1829 acting as Commissary-General of Prisoners during the Civil War. Frequently blunt and overly stubborn, he held deep racial prejudices. Hoffman not only discriminated in his treatment of the Tenth, but showed open contempt for the Black soldiers and their officers. His insistence that Black troops be kept at least ten yards away from the white troops and not be allowed to parade led to an open confrontation with Colonel Grierson. Despite Grierson's protests,

18

Colonel Hoffman also insisted on quartering the Tenth's recruits on low ground that filled like a swamp after every rainfall and he refused to provide for any type of walkways. These conditions led to widespread illness with many of the soldiers hospitalized for pneumonia. In July 1867, a cholera outbreak claimed sixteen of the new Tenth Cavalry recruits. Eight of those had been assigned to Company H. All eight died during the last week of July, causing additional reorganization within the company. Despite these setbacks, Carpenter's company was prepared to take the field in August 1867.[20]

Only a couple of weeks after Company H received their initial orders, another company of the Tenth Cavalry saw the regiment's first action against hostile Indians. The Cheyenne, who had been promised lands in Kansas and Texas under the Treaty of the Little Arkansas, struck back against expanding settlement when it became obvious that neither state would honor the terms of the treaty. Cheyenne war parties attacked railroad construction sites they believed were crossing their hunting grounds. This proved to be the "Indian war" Reuben Waller had heard about when he enlisted. On August 2, 1867, Company F under Captain George Armes followed the trail of Cheyenne warriors believed to have raided a work camp on the Kansas Pacific Railroad near Fort Hays, Kansas, killing seven workers. About eighty warriors attacked Armes' company in the field near the Saline River. After six hours, the Indians finally withdrew leaving the Tenth Cavalry with its first casualty, Sergeant William Christy of Pennsylvania, dead of a gunshot to the head.[21]

During those first months of duty, Company H troopers spent much of their time scouting, guarding the mail, protecting wagon trains and patrolling along the ever-expanding railway lines to prevent further attacks. In the latter part of August, the troopers rode west from Fort Leavenworth to Fort Harker in central Kansas. Colonel Grierson had managed to transfer his headquarters from the unpleasant conditions at Leavenworth to Fort Riley, stationing some of his companies at other posts throughout Kansas. When Company H stopped at Fort Riley along their march, Carpenter suffered his first desertion. Manuel Jones slipped away from the column on August 13. On August 20, they reached Fort Harker where Reuben Waller had his first sight of Indians. He later recalled, "The Indians at that time had overrun all of western Kansas and were robbing and murdering." Even though Waller had been a slave, he apparently did not sympathize or feel any kinship with the Cheyenne. Instead he felt appalled by what he saw and heard, although he was amused when the Indians seemed hesitant to fight or scalp the Black soldiers. He remembered that about a month after their arrival, five enlisted men and a white scout were captured while hunting buffalo. The Indians scalped and burned the white man alive at the stake, but they stripped and beat the Blacks and then sent them away supposedly saying "Indians no fight tashi-ti-bo-buthano" which Waller insisted meant "black man's scalp no good."[22]

Immediately after arriving at Fort Harker, Company H took the field, patrolling along the route of the Union Pacific Railroad. They established a camp on Walker's Creek at the end of the month. In his report, Carpenter wrote that through the month of October "The company was engaged guarding a force of workmen on the line of the Union Pacific Rail Road Eastern Division in Kansas." For this period, the captain rated his troopers excellent in military appearance, excellent at arms, good in discipline, instruction, clothing and accoutrements. While their service and appearance remained professional, field duty did prove very hard on the health of the soldiers. Henry Harper died of an unidentified disease on October 27, while recruit Perry Hogins' death in September was attributed to cholera.[23]

In November, Company H rode further west along the railway line until they reached Fort Hays. Then they turned east and reached the Tenth Cavalry headquarters at Fort Riley on December 3. Colonel Grierson had begun moving some of his troops into Indian Territory beginning with Company M assigned to Fort Gibson in November. While Carpenter's company had been in the field guarding the railroad, a peace commission had met in October with leaders of the Kiowa, Kiowa-Apache, Comanche, Cheyenne, and Arapaho on the Medicine Lodge River in Kansas to negotiate a treaty. The treaty resulted in a reservation being set aside for those tribes in the Indian Territory located west from the ninety-eighth meridian between the Washita River on the north and the North Fork of the Red River on the south. Stock theft and the illegal liquor trade already had become major problems for the tribes moving onto reservation lands and while it proved difficult to effectively stop those activities, the Tenth did establish regular patrols throughout the region. While at Fort Riley, two more of Carpenter's men deserted, John Williams and Charles Watkins. Duty in the field had presented hardships for which many of the recent recruits were not prepared. Even Civil War veterans found frontier duty more difficult than expected. Those veterans included Corporal Thomas Hayden and Sidney Sanders who deserted along with Ezariah Freeman and James Wright on February 29, 1868.[24]

Eventually desertion would become a rare occurrence for the men of Company H, but in the early days dissatisfied recruits attempted to escape and return to civilian life at the rate of about two per month. Of the eighty-four men recruited in 1867, twenty deserted during their first enlistment. John Thompson and Butler Tillman each deserted twice and Charles Shavers deserted on three separate occasions before he could be convicted, dishonorably discharged and imprisoned. While that level of desertion in Company H seems high, during roughly the same period of time, the Seventh Cavalry averaged 70 desertions per company or 3.5 per month. Some regiments reported as many as half of their soldiers deserting in a year. In the Black cavalry, the rate never reached that level, perhaps because the troopers placed a higher value on the opportunities the army provided or simply because there was no attractive avenue of escape near most of the posts where they served. Desertions appear to have been worst just before soldiers left the post on patrol or just after they returned from duty in the field. In May 1868, when Carpenter's troopers left Fort Hays marching west to Fort Wallace, Franklin Wilson saw his opportunity to desert. He boarded the eastbound train at nearby Hays City hoping to disappear into the crowd of travelers. Unfortunately for him, Major John E. Yard of the Tenth had a patrol at the station and Wilson was apprehended along with a deserter from another company. Private Wilson served one year at hard labor before the rest of his sentence was set aside. Sentences for desertion frequently depended on the circumstances and the judge, although they usually included a dishonorable discharge and at least one year in prison. Even that standard sometimes changed due to the circumstances. In August 1868, while Company H scouted near Fort Wallace, Private Irwin Boon deserted and rode south into Indian Territory. Three days later, Corporal Wall of the Fifth Infantry discovered Boon, still in uniform, at Hugo Wells. Recognizing a strange trooper who did not belong to his command, Wall arrested him at once. A court martial sentenced Boon to four months confinement and forfeiture of $10 per month for the same period.[25]

In May 1868, Company H settled into new quarters at Fort Wallace, the most western post in Kansas. Writing about the Kansas outpost during this period, the *Army-Navy Journal* reported, "The 'fort' is far from completion. The officers live in frame constructions, run up hastily from the ground, without foundation, and banked up, tent fashion, around the bases to keep out the wind." While many of the post buildings had been completed by this period, most of them did not even have doors. Company H did not have to be concerned about their living conditions on post for long. Soon after arrival at Fort Wallace, they joined five other companies of the Tenth Cavalry in a scouting expedition against the Cheyenne. The Medicine Lodge treaty not only had provided reservation lands, but also had made promises of food and clothing, weapons and ammunition, and the right to hunt buffalo south of the Arkansas River. When Congress failed to implement the provisions of the treaty and provide expected supplies, raids were renewed throughout Kansas, Colorado and Texas. Some of the Kiowa and Comanche bands had come to the Wichita Agency near Fort Cobb to obtain supplies from the agent that spring, only to find that no supplies had arrived. Seeing this evidence that the new treaty had not been kept, they drove out the agent, destroyed the agency building, and left the reservation, returning to the upper area of the Arkansas River along with groups of the Cheyennes and Arapahoes. Colonel Grierson moved his headquarters to Fort Gibson in May 1868 to more effectively combat those Kiowa and Comanche raiders who had attacked the Wichita agency, threatened the Chickasaw reservation and even begun to raid south into Texas. He sent Major Meredith H. Kidd with six companies into the field near Fort Wallace to guard against further Cheyenne raids in Kansas.[26]

Major Kidd's detachment marched out of Fort Wallace along the North Fork of the Smoky Hill River scouting an area of over 100 miles along the Denver stage road during June. In July they scouted as far north as the headwaters of Eagle Tail Creek and established camp near the Schimmerhorn Ranch on the Republican River. While they had yet to encounter any Cheyenne raiding parties, on July 14, they did receive a message from Tenth Cavalry headquarters warning that, "Cheyennes left their camp yesterday morning, moving north. They are not contented with the refusal to let them have arms, and some of them may possibly do mischief, therefore commanders will be on their guard for fear of a surprise." Throughout August troopers of the Tenth Cavalry continued on fruitless patrols covering more than one thousand miles back and forth across western Kansas.[27]

With the prevalent threat of Cheyenne reprisals, Company H left Major Kidd's camp on detached service during August. Riding to Spellman's Creek near the Saline River, they "assisted in building a Blockhouse for the protection of settlers in the vicinity." The isolated communities in that area had heard of raids on settlements along the nearby Solomon and Smoky Hill routes of the Denver stage road. To provide some measure of security, the troopers remained with them for several weeks instructing the settlers concerning their defense and helping to fortify the farmers and ranchers against future raids.[28]

Company H returned to Fort Wallace on September 17, 1868. They now had been in active service on the Kansas frontier for over a year. The Black troopers were no longer raw recruits, having spent well over half of that first year on scouting patrols. Already the reputation of the Buffalo soldier cavalry had begun to grow. A writer for Harper's Weekly noted that although many ranchers had been skeptical of the troopers' abilities at first, now he could confidently say, "I have not met a single frontiersman who has seen the dusky patriots 'go for Indians' but is loud in their praise." Carpenter's company had yet to "go for Indians," but there would be opportunity with the Cheyenne on the move again. The long-awaited Medicine Lodge Treaty supplies, including guns and ammunition, had begun to arrive in August and many of the war parties began to move toward reservation lands in Indian Territory. In the coming weeks of September and October 1868, Company H would take the field again with a chance to demonstrate the value of their first year's training.[29]

RIDING TO THE RESCUE:
KANSAS AND INDIAN TERRITORY 1868-1871

Private Reuben A. Waller
Courtesy Kansas Historical Society.

The year 1868 marked many changes for the frontier army. Threats along the Mexican border had diminished following the demise of Emperor Maximilian in June 1867. The number of troops occupying the former Confederate states gradually decreased as those states began to re-enter the Union with new state constitutions. Reductions in force had already begun that would reduce the peacetime army from 57,000 men in 1867 to about 26,000 ten years later. The eastern states, weary of war, seemed anxious to make peace with the Plains Indians rather than trying to defeat them. The army's task would be to enforce that peace. In August 1868, the government finally began to implement the provisions of the Medicine Lodge Creek treaty, distributing provisions to the Cheyenne, Kiowa, and Comanche on the reservations in Indian Territory. Unfortunately, these provisions included an ample supply of guns and ammunition intended for hunting. When the Indians used these guns for raids against settlers, the army attempted to drive all remaining Indians out of Kansas and south onto reservation lands. On August 10, 1868 a party of over 200 Lakotas, Cheyennes and Arapahoes attacked settlers in the Solomon and Saline Valleys of central Kansas. The Indian raiders killed thirteen and carried off one woman captive. Seven companies of the Tenth, including Company H, marched to the aid of the settlers, working with them to construct blockhouses for their defense.

In an effort to track the raiders, on August 24, Major General Philip H. Sheridan appointed his former aide from the Civil War campaign in the Shenandoah Valley, Major George A. Forsyth of the Ninth Cavalry to recruit a company of fifty civilian scouts. Called by some the Solomon Avengers, they assembled along the railroad line in Kansas at Fort Hays, Fort Harker, and Fort Wallace. One of the men recruiting scouts for Forsyth was James J. Peate, better known as Jack. Peate, the nineteen-year-old son of a Methodist minister, had run away from his home in Ohio two years earlier. He later claimed to have recruited thirty men for Forsyth's command, but only fifteen reported "on account of the bad weather." The Solomon Avengers set out from Fort Wallace on September 10 trailing a raiding party that had attacked a wagon train at the railhead of the Kansas Pacific Railroad killing two teamsters. Peate remained behind on duty at the post. They picked up the party's tracks at the site of the recent attack near Sheridan, Kansas. Forsyth persistently followed the trail into a broad valley on the Arickaree Fork and camped across from a small, narrow island. At dawn on September 17, a combined force of Cheyennes, Arapahoes and Sioux attacked. The scouts quickly took refuge on the island, digging rifle pits as over 600 Indians surrounded them. Forsyth's lieutenant, Frederick Beecher and Dr. John Mooers received wounds during the early attacks from which they later died as did the Cheyenne leader, Roman Nose. The major sent scouts Jack Stilwell and Pierre Trudeau to Fort Wallace for help. Two days later, he sent Allison Pliley and Jack Donovan. The rest of the command remained pinned down on the site they had now begun to call Beecher's Island, after the dead lieutenant.[30]

Unaware of Major Forsyth's situation, Captain Carpenter left Fort Wallace on September 21 with sixty-nine troopers of Company H and seventeen scouts to patrol the Denver road and keep it clear of Cheyenne raiders who were "seriously molesting the stages and interfering with the delivery of the mails." Thirteen wagons carried thirty days rations for the men of Company H. On the morning of the 23rd, two couriers reached Carpenter's camp with the news of Forsyth's company and orders from Captain Henry C. Bankhead, post commander at Fort Wallace. Company H was to proceed immediately to the Dry Fork of the Republican River to aid the besieged scouts. Carpenter clearly remembered Forsyth from their service together on the General Sheridan's staff in the Shenandoah Valley. Determined to relieve his old comrade, the captain left the stage road with his supply wagons and began the long ride across the plains. He later wrote, "In the lowlands there ran a large fresh trail, over which at least 2000 head of ponies had recently been ridden or driven down the stream. It was so fresh that I was apprehensive that the Indians were near at hand, and therefore pushed on rapidly to the side of the stream, where the wagons were corralled and preparations made for a possible conflict."[31]

Beside Carpenter rode Private Rueben Waller in his current duty as the captain's orderly. With the company scouts was Jack Peate, anxious to find and rescue the men he had helped recruit and send into danger. Peate later recalled, "Not a man, horse, or mule, but did all that was required of him." Also riding with the troop was Doctor Jenkins Fitzgerald, an army surgeon and a Civil War veteran from Indiana. Although only 29 years of age, Doctor Fitzgerald had extensive battlefield experience and had been described by his commander during the war as "manifesting a thorough disregard of his own safety in his humane desire to give the wounded the promptest surgical relief." In his orders, Captain Bankhead had asked Carpenter to send the doctor back to Fort Wallace before proceeding up the Republican River. Instead, Carpenter decided to keep Fitzgerald and his ambulance with the troop, knowing Forsyth would probably have many sick or wounded men.[32]

By the afternoon of September 24, the troopers reached what they thought was the "Dry Fork" of the Republican, yet they found no evidence of a battle and no scouts. Captain Carpenter climbed a hill to get his bearings and discovered Indian burial scaffolding indicating a number of recent burials. On further inspection, the soldiers found fresh bullet wounds in many of the bodies. "On the opposite side of the river and up a small ravine," wrote Carpenter, "we found a small teepee of clean white robes, and on the frame inside lay the body of a warrior wrapped in buffalo robes." It was the Cheyenne chief, Roman Nose. Certain that Forsyth must be nearby, Carpenter sent out scouts to follow the Cheyenne trail from the burial site. Before they departed, a party accompanying Jack Donovan, one of Forsyth's scouts who had gone for help, rode into Carpenter's camp. Following Donovan's directions, Carpenter took a detachment of about thirty men and "a light ambulance" forward riding through "the rough and rugged breaks." At about 10:00 a.m. on September 25, they located the battle site.[33]

At Beecher's Island, Forsyth's men had begun their ninth day, not knowing if they could even dare to move from the sandy rifle pits. Scout Chauncey Belden Whitney recalled, "About ten o'clock the cry of Indians rang through the works. Some of the men being out, eight or ten of us took our guns to rescue them if possible. The word was given that it was 'friends.' In a few moments, sure enough, our friends did come. Oh the unspeakable joy." One of the scouts looking up shouted, "I think that's Doctor Fitzgerald's greyhound." Another cried out, "By the God above us, it's an ambulance." In a few moments, Captain Carpenter, with Reuben Waller beside him, rode up and dismounted just in front of Forsyth. Waller later wrote, "They ought to remember Col. Carpenter and his colored orderly, as we charged up to where they were lying in their rifle pits and how we all cried together as we helped them out of their starving condition." Major Forsyth, retaining his dignity as he greeted his old comrade, remembered, "When Colonel Carpenter rode up to me, as I lay half covered with sand in my rifle-pit, I affected to be reading an old novel that one of the men had found in a saddle pocket. It was only affectation..." He continued, "Strong men grasped hands, and then flung their arms around each other, and laughed and cried, and fairly danced and shouted again in glad relief of their long pent-up feelings."[34]

Waller and the other troopers were shocked at the sight of wounded scouts half buried in the sand with forty-seven dead horses and mules around them. "And these men had eaten the putrid flesh of those dead horses for eight days." Waller declared, "We began to feed the men from our haversacks." Doctor Fitzgerald had to stop the soldiers from overfeeding the scouts and making them even sicker. Captain Carpenter "pitched some tents a quarter of a mile farther up the stream, carried all of the wounded to the new camp, and made them as comfortable as possible." Forsyth proved to be "too weak, shattered and nervous to be able to talk much." His leg had been splintered and blood poisoning had set in. Doctor Fitzgerald went to work, treating the wounds of Forsyth and his men. Carpenter later credited him with saving Forsyth's life and leg. On September 26, Captain Bankhead arrived with additional troopers from Fort Wallace. The soldiers remained on the island caring for Forsyth's scouts for several days. They buried Lieutenant Beecher and four others with military honors using Company H's funeral flag. Sixty years later, Reuben Waller still kept that flag as a prized possession.[35]

"The Rescue" - Captain Carpenter and Major Forsyth at Beecher's Island. Reprinted from *Harper's New Monthly*, June 1895, page 58.

On the morning of September 27, troopers and ambulances left Beecher's Island for Fort Wallace. That night they camped on the south branch of the Republican where scout Chauncey Whitney recalled, "Five of our boys killed and scalped a Cheyenne about one-half mile from camp." Company H arrived at Fort Wallace on the evening of September 29 while Captain Bankhead remained encamped with many of the more seriously wounded a few miles from post. Sergeant John D. Price of Company H ordered Blacksmith Ephraim Smith to return to Bankhead's camp with an ambulance and help bring the wounded back to the post. Instead, Smith decided to leave the ambulance and visit the saloons of nearby Pond City. Smith remained in town all night. Court-martial charges against him claimed that while "under the influence of liquor (he) discharged his carbine into the town several times, thereby endangering the lives of several citizens and creating great alarm and excitement." Ephraim Smith had seen the aftermath of battle for the first time that week and his actions were probably understandable. He was found guilty and fined ten dollars for one month. Captain Bankhead's command brought the rest of the wounded into the hospital on September 30.[36]

About two weeks later according to Reuben Waller, some of the soldiers took "French leave" and met scouts of Forsyth's command at Pond Creek three miles west of Fort Wallace to celebrate and got "very wet drunk." "They sure treated us black soldiers right for what we had done for them," Waller exclaimed. But Waller and the other troopers did not have much time to settle into the routine of life at Fort Wallace and enjoy leave, official or unofficial. A well-marked Indian trail had been discovered leading toward Beaver Creek during the Beecher Island rescue. The Fifth Cavalry, en route from Fort Hayes to Fort Wallace by rail received orders to disembark and follow the trail. There was only one problem with that order. The new commander of the Fifth Cavalry, Medal of Honor recipient, Colonel Eugene A. Carr, had already arrived at Fort Wallace and was awaiting his regiment. Since General Sheridan wanted Carr, an experienced commander, at the head of his troopers, he ordered an escort to take the colonel to Beaver Creek. Captain Bankhead offered Sheridan Companies H and I of the Tenth Cavalry under Captain Carpenter as escort.[37]

Up to that point Carr had been among those who were less than favorable toward the idea of using Black troops on the frontier. He insisting on observing the two companies as they drilled in marksmanship and horsemanship. Reuben Waller later claimed that Carr was so impressed by the Black soldiers' drill that he told the post commander, "Why, Bankhead, I can take those two troops of negroes and whip hell out of all of the Indians in Colorado." Captain Bankhead replied, "Very well General Carr, they are at

29

your service." At about 10:00 a.m. on October 14, 1868, Captain Carpenter's column left Fort Wallace accompanied by eleven wagons loaded with campaign supplies and forage for the horses. Carpenter recalled, "The mules, dragging heavy over rough country, were made to trot in order to keep up with the cavalry column." On October 15, they reached Beaver Creek and proceeded down the stream. After sixty miles march along the creek, they had found evidence of one Indian camp but no sign of the Fifth Cavalry. Carr began to wonder if his unit had ever made it as far as Beaver Creek.[38]

At about 7:00 a.m. on October 18, while scouting to the front of the column, Captain George Graham of Company I encountered twenty-five Indians who attempted to cut him off. Carpenter saw the attack in progress and ordered thirty men forward to Graham's relief. At the same time he led his troops across the creek and formed the wagons in double columns to create as compact a formation as possible. Company H formed rifle lines along the flanks and in advance of the wagons while Company I covered the rear. In this formation, they began to advance "steadily up the creek bottom" while the Cheyennes harassed the column from any available cover. Carpenter attempted to move his train to higher ground, but as he later remembered, "At one point we passed near a deep ravine, and the enemy, quick to observe cover of any kind, occupied it with quite a number of warriors and opened up a serious fire." Captain Graham then led the reserve platoon to attack the Indians in the ravine while the wagons continued to move on through along the creek. Carpenter observing the advancing Cheyennes noted, "They were all stripped to the waists, and were decorated by various ornaments hanging from their heads and shields, quivers and bridles, so as to glisten and shine in the sun at every turn of the ponies."[39]

At about 1:00 p.m. the Indians withdrew. Half an hour later they reappeared, 400 to 700 strong, prepared for a full assault. Carpenter offered command of the cavalry to Colonel Carr, but the colonel refused. Carr later claimed he had always been in command and had simply let Carpenter take action with his troops as long as his orders remained sound. Carpenter's next order was to form the wagons into a horseshoe shaped corral with the mules facing inward and the horses tied together inside the corral. All this was completed in about two minutes, while the troopers built breastworks out of forage sacks and cracker boxes. As the Indians charged and circled, the buffalo soldiers fired their new seven-shot Spencer carbines at the Cheyenne warriors who were riding eight to ten deep. Accurate fire from the trial-use Spencers kept the Cheyennes at a distance. As the first wave of Indians retreated, a warrior Carpenter took to be a chief medicine man rode out about 200 yards away "to try to show that his medicine was good and the white man's bullets could not hurt him." The carbines of Company H quickly cut him down. Reuben Waller later claimed that the troopers "being dead shots, the Indians melted away like snow beneath the burning rays of a summer's sun. We slaughtered them unmercifully." According to Waller, Colonel Carr praised the soldiers' marksmanship claiming, "Men you have surely gained the day." As the Cheyennes moved away, some of the troopers began to scalp the dead Indians, said Waller, "but we soon put a stop to that kind of barbarity among the tenth cavalrymen." In all, nine Indians lay dead and about thirty wounded. Three soldiers had been wounded including Private John Claggett who had been hit in the leg with an arrow.[40]

Late in the hot afternoon, Carpenter decided to move back to Beaver Creek for water. With the wagons again in double columns, horses on the inside, men dismounted on the outside, they slowly returned to the stream. That night the soldiers camped on a wide creek bottom, digging rifle pits for their pickets. "We heard them around us all night imitating coyotes," Carpenter remembered, but the Cheyennes never found a weak spot to break through. On October 21, Companies H and I returned to Fort Wallace after riding 230 miles without finding the Fifth Cavalry. Along their march, the soldiers passed through Sheridan City, railhead of Union Pacific Railroad. As they rode down the main street, men poured out of the numerous saloons and as Carpenter put it "when they saw the troopers and their horses decorated with the spoils from the Indians whose dead bodies we had captured, they knew that we had been in a successful fight and they gave us a perfect ovation."[41]

During that period, Colonel Grierson had divided his command among various stations throughout the area to control Indian bands moving north and west. Companies B, F, G and K patrolled out of Fort Lyon in southeastern Colorado, while

Companies A, C, H and I patrolled along the southern Kansas border. Four companies had moved into Indian Territory, Companies D, L and M at Fort Cobb and Company E at Fort Arbuckle. During the winter of 1868-1869, General Sheridan decided to launch an all-out campaign to force the Cheyennes, Arapahoes, Kiowas and Comanches onto reservations in the western half of the Indian territory. Three columns set out to trap the Indians between them. The main column left Fort Dodge under Lieutenant Colonel Alfred Sully with eleven companies of the Seventh Cavalry and five companies of infantry marching along the Beaver River. The second column set out from Fort Lyon, Colorado in two separate forces. The first to leave consisted of the four companies of the Tenth Cavalry already stationed there and the remaining company of the Seventh Cavalry commanded by Captain William H. Penrose. Almost immediately they were halted by a snowstorm and forced to take shelter and wait for the remainder of the column consisting of seven companies of the Fifth Cavalry under Colonel Carr with a large supply train and a herd of cattle. The third column from Fort Bascom, New Mexico was commanded by Major Andrew W. Evans with six companies of cavalry, two companies of infantry and a battery of mountain howitzers. They marched east along the Canadian River through blizzard conditions. On Christmas day, Evans' command attacked and destroyed a Comanche village at Soldier Spring in the Texas Panhandle. In November, Colonel Sully's column, also delayed by the harsh winter, established a base at the junction of the North Canadian River and Wolf Creek in Indian Territory that they named Camp Supply. While there, animosity increased between Sully and his second in command, Lieutenant Colonel George Custer. Custer believed he should be in command as both were actually lieutenant colonels and Custer held a higher brevet rank. General Sheridan solved that conflict when he visited Camp Supply on November 21. The general ordered Sully to return to Fort Harker, Kansas to take up duties as commander of the Upper Arkansas District leaving Custer in field command. Marching through sleet and snow, Colonel Custer's Seventh Cavalry attacked Black Kettle's Cheyenne village along the Washita River on November 27. Custer had correctly determined that hostile Indians were camped in the area of the Washita; however, Black Kettle's band was not among those renegades. After surviving the massacre at Sand Creek in 1864, Black Kettle had attempted to remain at peace with the white settlers. This time Black Kettle did not survive his continually frustrated efforts to maintain peace. He died at the hands of Custer's men along with many of his village including women and children. As the soldiers advanced, warriors from some of the neighboring villages who were actually hostile to the soldiers counterattacked and Custer was forced to withdraw in the darkness. General Sheridan returned to the Washita with a large force from Camp Supply on December 7. His soldiers followed the fleeing Kiowas and Comanches to Fort Cobb where the Indian agent, General William B. Hazen, informed Sheridan that they had surrendered to the Bureau of Indian Affairs. Sheridan immediately arrested Kiowa chiefs, Satanta and Lone Wolf, and threatened to hang them if all of the Kiowas did not surrender. By Christmas, only a few Cheyenne and Arapaho bands remained absent from the reservation.[42]

Company H took the field again on November 20 to scout south of the Arkansas River with Company I, Tenth Cavalry and Company I, Thirty-eighth Infantry. According to Captain Carpenter "No hostile savages were encountered," but on December 5 a severe blizzard struck the column. With no cover anywhere in sight, Carpenter continued on "marching thirty miles in the face of the storm to partial shelter at Monument Station on the stage road. Many animals were frozen or died from exhaustion, and a number of men were forced afterwards to have portions of their feet amputated in the hospital." About thirty mules and several horses froze to death. Although the soldiers used parts of the wagons for firewood in an attempt to stay warm, at least sixteen members of Company H were treated for frostbite on their return to Fort Wallace. Among those suffering the loss of toes were Silas Jones, Alfred Owings, Simon Peter and John Billings. James Clayton, James Thomas, Scott Bridgewater, and John and Joseph Claggett also received treatment for their frozen extremities. In later years the blizzard of '68 became a favorite story whenever veterans of Company H gathered to remember the harsh conditions they had survived in frontier service.[43]

On January 21, 1869, a correspondent at Fort Wallace wrote in the *Army-Navy Journal*: "An expedition left this post today to scour the country from Sandy to the Republican in search of Indians." The column under Captain Bankhead's command included Companies A, H, and I, Tenth Cavalry and Companies B and I, Fifth Infantry with rations for twenty-five days and a wagon train of seventy-five wagons. The night before they departed, strained relations between Black and white troopers led to some members of Company I, Fifth Infantry firing their rifles wildly into the barracks of Company I, Tenth Cavalry, wounding three Black soldiers. Four of the Fifth Infantry privates were arrested and charged for the incident. Confined, but not convicted, one of the white soldiers, Albert Horton, deserted soon after being released. The *Journal* correspondent noted, "The white and colored troops had a small row last night, which resulted in three colored men being placed hors de combat. One was wounded in the left arm, which was amputated by Dr. Fitzgerald in artistic style. The other two were wounded in the legs, and it is a question whether they do not loose a leg apiece." That regrettable incident between white and Black troops on the eve of their march proved to be the only action seen by Bankhead's patrol. After what must have been an uneasy march, the weary soldiers returned to the post on February 12 without sighting a single Indian.[44]

In the spring of 1869, Colonel Grierson became engaged in establishing a permanent reservation for the Comanches, Kiowas, Cheyennes, and Arapahoes near the Wichita Mountains on Medicine Bluff Creek. His headquarters, originally known as Camp Wichita, stood at the junction of Medicine Bluff and Cache Creeks. That outpost was officially designated Fort Sill in August 1869. Through most of the spring, Company H had remained in Kansas at Fort Wallace. Alexander Adams had died there on April 9 of acute rheumatism. Many of the troopers suffered from scurvy after frequent cold weather scouting expeditions with little opportunity to recover. On April 8, General Hazen, the Indian agent at Medicine Bluff Creek wrote that Kiowas and Comanches "who have been accustomed to follow the buffalo north in the spring are talking of doing so this season as formerly" thinking they would receive food at Camp Supply and other posts along the Arkansas River. He requested that no rations be issued to any Indians who had left the reservation. To guard against any troubles that might arise through enforcing this policy, Companies H and I departed for Camp Supply on May 19 under the command of Lieutenant Colonel Anderson D. Nelson.[45]

Camp Supply along the Canadian River in the northwestern corner of Indian Territory had been established the previous winter to support General Sheridan's campaign. Forrestine "Birdie" Cooper, daughter of Lieutenant Charles Cooper, wrote her first impressions of the outpost during this period, "A terrific thunderstorm was raging when our ambulance reached Camp Supply late the next afternoon. We were driven past a long row of squatty log cabins that were connected at the sides by small porches about four feet across, giving the effect of one continuous building. There were no doors to be seen at the rear." Officers' Row proved to be a collection of picket houses with dirt floors. Cooper later commented that the covered side porches could be used by the soldiers to fire from cover while the women and children retreated from house to house to escape Indian raids. She continued, "The dust storms at Camp Supply were frequent and fierce. The fine sand sifted in clouds through the house and filled the cooking pots. Often it was necessary to use candles in mid-day, for coal oil was too valuable for constant use even when it could be obtained from a venturesome peddler. I recall a different type of storm at Camp Supply—a grasshopper storm. The insects came in a dense cloud, like hailstones, and fell around our doors and covered the ground." By the summer of 1869, Camp Supply had become home not only to dust and grasshoppers, but also to about 1320 Arapahoes living in 257 lodges around the post.[46]

Since a permanent reservation with a civilian Indian agent had not been established, one of the first tasks facing Company H and others of Colonel Nelson's command was the distribution of food and supplies to the Arapahoes. Reuben Waller exclaimed, "Here we had six thousand Indians on our hands to feed." Soldiers unloaded rations from the wagons and separated them into piles. Women and children sat in a large circle around the rations, while the warriors formed a smaller circle inside them. The "young chiefs" distributed each item until every family had received their share and then they returned to their lodges. Concerning the Arapahoes' behavior, Colonel Nelson stated, "The people are entirely friendly and desire to remain at peace and a little kind treatment will secure peaceful relations with them for the future."

Waller also recalled the troopers herding cattle to feed the Arapahoes.[47]

The move from frontier settlements in Kansas to the remoteness of Camp Supply caused morale to reach one of its lowest points since the organization of Company H two years before. Captain Carpenter had been absent on detached service and leave since April and had not been part of the move to Camp Supply. Lieutenant Louis Orleman was absent most of the summer as a court martial witness, leaving Lieutenant Charles Banzhof as the only company officer present. Part of the summer, Banzhof also served as battalion adjutant. Banzhof, like Orleman had immigrated from Prussia before the Civil War and may have had some difficulty communicating with the Black troopers. In addition to difficult living and working conditions, Banzhof suffered from chronic diarrhea and it was rumored that he sometimes drank heavily. This left much of the organization and administration of the company to the First Sergeant, Amos Cormack. Sergeant Cormack may have been a poor choice for command as he had not been known to enforce discipline even before the move to Camp Supply. In January 1869, he was accused of allowing soldiers passes to visit the saloons of Pond City near Fort Wallace after Captain Carpenter had specifically ordered the town placed off limits. With such command problems added to the continuing hardships of life at camp, the lure of escape from a isolated outpost proved too tempting to resist.[48]

On August 6, 1869, ten members of Company H deserted, the worst incidence of desertion up to that time, as contrasted to their exceptional record in the field. While the records are not complete, it appears that most of the deserters were on guard duty and abandoned their posts, taking their weapons and equipment. First Sergeant Amos Cormack and Alfred Dixon, the corporal in charge of the stable guard, both deserted their duties along with Privates Richard Gowans, William Johnson, Allen McPherson, David Mead, William Oliver, Charles Shavers, James Thistle and James Wright. Some of the deserters had prior charges against them. Shavers had been accused of brandishing a revolver at a prisoner in the guardhouse and Wright was charged with stealing blankets from a local teamster. The desertion of senior enlisted men with full arms, horses and equipment constituted a serious violation of duty and a thirty-dollar bounty was offered for the capture of each man. Oliver and Johnson were both caught within a matter of days. Each received a prison sentence and a dishonorable discharge. Army patrols captured Shavers on August 12, but he escaped again on August 17. Corporal Dixon finally surrendered himself in October. He received a reduction in rank, a dishonorable discharge and four years confinement at Jefferson City, Missouri. There is no record that the other six deserters were ever found.[49]

Captain Carpenter returned from leave on August 28. Under his direction, the troopers, who had been living in tents, set to work building temporary barracks and corrals for the coming winter. By the end of October 1869, Company H had only thirty-nine men either present or on extra duty. To fill out the ranks, twenty-five new recruits arrived on December 8, including Henry Allen, Lewis Hayes, Louis Mack, James Wilson and Anderson Wilson. Most of the new recruits continued to be from border states such as Missouri, Arkansas and Kentucky. They were all in their early twenties and many had been farmers or common laborers. Recruits for white regiments at that time appear to have been of similar age and occupation, but most were either foreign born or came from northern and midwestern states. With the stabilization of command and filling out the ranks, conditions on post did improve, yet desertions continued to occur. In April 1870, three more troopers abandoned the post including Joseph Forrest, who had arrived with the recruits in December. A few months later, recent recruit Lewis Hayes deserted along with Robert Edwards, William Hamlin, John Henry and William Ross. All received dishonorable discharges and three years confinement.[50]

Although the Arapahoes seemed to have settled into reservation life in 1869, the same could not be said for the Cheyennes, Comanches and Kiowas. Reuben Waller recalled a Cheyenne attack on a nearby stage station in which eight white infantrymen were killed in September 1869. Waller helped bury the soldiers several miles north of Camp Supply, sadly commenting "just as we were getting acquainted" with the Indians, they had gone on the "war path." Captain Carpenter remembered that Company H spent much of 1870 patrolling the country around the reservations and escorting wagon trains, due to "the Cheyennes becoming hostile." Early on the morning of May 30, 1870, two teamsters slipped away from a train of thirteen wagons loaded with stores for the Indians around Camp Supply. Kiowas had attacked the train two days earlier along the North Fork of the Canadian River, stampeding fifty-eight mules and killing one of the teamsters. When the two men reached Camp Supply later that day, Carpenter immediately mounted Companies H and I and rode out to relieve the besieged train. By the time Carpenter arrived Lieutenant Mason Maxon had already reached the train with four more wagons. The troopers drove away almost 300 Kiowas and then escorted all seventeen wagons to Camp Supply. A few of the frustrated Kiowas killed a lone teamster about three miles from the post and scalped, stripped and mutilated him.[51]

Less than two weeks later, on June 10, Carpenter's troops again left camp in force to intercept an oxen drawn train bound for Fort Dodge, that was under attack along Snake Creek. The captain reported that they reached the train in good time and the Indians "were repulsed with no loss." They had barely returned to post when a large party of Comanches attacked Camp Supply on June 11, with the intention of stampeding and capturing the horses on the picket line. The Comanches' intent had been to conduct a swift raid before the soldiers could respond, but five companies of the Tenth Cavalry retaliated almost immediately, saving most of the horses. They continued to skirmish with the dismayed Indians for over an hour. When the Comanches finally retreated, they left six dead and ten wounded. Three troopers of the Tenth had sustained wounds during the fight.[52]

On October 15, 1870, Colonel Nelson reported that "Company H left Supply this day" for their new station at the Tenth Cavalry headquarters, Fort Sill located at the junction of Cache Creek and Medicine Bluff Creek against the backdrop of the Wichita Mountains. At that time, he also noted that the Cheyenne and Arapahoes were no longer satisfied with their location because of the scarcity of buffalo, signifying the probability of future Indian problems. Company H arrived at Fort Sill on October 28 and quickly became involved in Colonel Grierson's construction program working through the winter on a new barracks, stables and corrals to combat the problem of overcrowding. Spring brought the anticipated Indian troubles or as Reuben Waller said, "Now came some hard times for us worn out soldiers." On May 12, 1871,

Kiowas and Comanches ran off some stock near the Red River. Troopers from Fort Sill pursued the renegades killing three and wounding four.[53]

Then on May 19, one wounded teamster stumbled into Fort Richardson, Texas. The Warren train of ten wagons carrying supplies to Fort Griffin had been attacked by Kiowas from the Medicine Bluff Creek reservation. Seven teamsters had been killed, some "bound to the wheels and burned to death" and forty-seven mules stolen. General William T. Sherman, passing through Fort Richardson on a tour of inspection, immediately ordered Colonel Ranald Mackenzie and his Fourth Cavalry to hunt down the Kiowas guilty of such an outrage. Sherman then continued to the next station on his tour, Fort Sill. There he discovered that the Kiowas had already returned to the reservation and as Reuben Waller put it, they were "all in camp playing 'good Indians.'" Indian agent Lawrie Tatum, described as an "honest old Quaker," questioned the Kiowa chiefs when they arrived at the agency for rations. Satanta, a tall, boastful man of about forty, proudly claimed that he had led the raid and he named the other chiefs involved. Those chiefs confirmed Satanta's claim. Sherman decided to call a council on the front porch of Colonel Grierson's quarters to arrest the chieftains responsible.[54]

Twenty troopers waited with carbines ready behind the shuttered windows of Grierson's home. Captain Carpenter with Company H and three other companies of cavalry waited armed and mounted within the high-walled corrals they had helped to build. When the Kiowa chiefs arrived, Sherman told them that those guilty of the wagon train raid would be arrested and taken back to Texas for trial. Satanta, who had bragged of leading the raid, grabbed for a pistol under his blanket. At a signal from Grierson the porch shutters flew open revealing an armed squad of soldiers. Carpenter and Captain Richard Pratt trotted their men out from either side of the corrals with weapons ready. Lieutenant Louis Orleman, with a detachment of ten men from each company, approached the watching Indians from the rear. At that point Lone Wolf arrived with two carbines and a bow and quiver of arrows. He handed the bow to one Kiowa, the extra carbine to another and stood facing the porch menacingly. A fight seemed inevitable when suddenly Grierson tackled Lone Wolf and held him to the ground shouting that violence would not save the guilty chiefs. After a few minutes of tension, Carpenter's men cleared the area of Kiowas and Comanches while Satanta, Satank and Big Tree were placed under arrest and prepared for transport back to Texas. On May 30, forty-one men of Company H left Fort Sill to escort General Sherman on the next leg of his tour toward Fort Gibson. The three Kiowa chiefs left under guard for Fort Richardson on June 8. Satank, a small, cruel looking man of about sixty, had concealed a knife. He managed to free himself and attacked his guard after less than a mile on the trail. Troopers riding up from behind shot him down and left his body at the side of the road. Satanta and Big Tree continued under guard to Jacksboro where they were tried by a civil court and convicted of murder.[55]

Alexander S. B. Keyes and Virginia Maxwell Keyes

During the same period, Lieutenant Alexander Keyes joined Company H as second in command. Keyes, born into a military family at Deadham, Massachusetts on July 28, 1846, joined the Union army in 1863 when he was only seventeen. When the Civil War ended he requested assignment to the regular army. In 1869, he was placed on detached duty to serve as Indian agent at the Apache and Ute reservation in Cimarron, New Mexico. There he met Virginia Maxwell. Virginia was the daughter of Lucien Maxwell, owner of one of the largest Mexican land grants ever issued. Keyes proposed to marry Virginia, but Maxwell had already promised her in marriage to a nearby rancher. Being a strong-willed young woman, Virginia plotted with the Methodist circuit rider, Thomas Harwood to wed Keyes in secret. March 30, 1870, was a mill day, when the Utes and Apaches came to the mill in Cimarron to receive their rations. Keyes, the Indian agent would be present and Virginia arranged to be at the mill as the guest of the miller, Isaiah Rinehardt and his wife. Late in the afternoon, Virginia Maxwell and Alexander Keyes were married by the Reverend Harwood on the third floor of the mill with the Rinehardts as the only witnesses. After the ceremony, the newlywed Keyes returned to their separate homes to await the lieutenant's anticipated orders to join the Tenth Cavalry.[56]

Keyes' orders to report for duty with the Tenth at Fort Sill arrived the first week of April and several weeks later the couple boarded a northbound stage. Once they had left the Maxwell land grant near Trinidad, Colorado, "Keyes handed the stage conductor a copy of the marriage certificate for delivery to Maxwell." Reverend Harwood decided it would be best to avoid Cimarron for a while. As he skirted around the settlement on his next rounds, he came upon an Apache camp. He recalled, "Cimarron and Maxwell in the rear and hostile Apaches in my front. What shall I do? Of the two dangers I concluded to choose the least, as I thought and pressed on" toward the Apaches. In reality, Lucien Maxwell posed no serious threat. Although angered at first, he soon forgave his favorite daughter. While Lieutenant Keyes rode patrols against the Kiowas in 1871, Virginia returned home to the Maxwell ranch in New Mexico to give birth to their first child. During this period, Lieutenant Banzhof requested to be discharged from the service and was replaced by Second Lieutenant William R. Harmon. Harmon was later described as "one of the most intrepid who ever exercised a commission in the 10th" and who "never required men to go into the jaws of death unless he was at the front."[57]

During the summer of 1871, Company H remained in the field, guarding against further Kiowa breakouts. On July 28, Reuben Waller suffered a gunshot wound to the side of his face. Evidently, the wound was accidental, but Waller continued to receive treatment through February of 1872. In August, Captain Carpenter commanded a battalion consisting of Companies G, H and I "in search of marauding Indians along the line of Red River."

They operated by sending regular patrols out of a camp on Otter Creek that Carpenter claimed had been "established to watch and overawe the Kiowas" who were known to have stolen forty-one mules. At the same time Colonel Mackenzie and the Fourth Cavalry patrolled the North Fork of the Red River on the Texas side. Traveling through rough, broken country in sweltering summer heat, neither column saw any action, although they did receive word while in the field that the Kiowas had returned the stolen mules. To ease the frustration Colonel Mackenzie must have felt, Carpenter had a surprise prepared for the Fourth when they reached Otter Creek in September. Despite the primitive camp conditions, Mackenzie and his staff were served a magnificent feast from Carpenter's field kitchen that included turkey, quail, buffalo, fish and even prune pie.[58]

As they returned to Camp Supply in the fall of 1871, many of the men of Company H were nearing the end of their first five-year enlistment. They had taken part in numerous patrols and skirmishes against the various tribes of the Southern plains throughout Kansas and Indian Territory. They had dealt with primitive living conditions, long patrols on horseback, harsh weather, and the endless routine of life on a frontier outpost. Sixteen of the troopers had been granted disability discharges due to injury or serious illness. Over twenty soldiers had deserted and at least three had received dishonorable discharges for causes other than desertion, including one for murder. Five troopers died during their enlistment, none of them from wounds suffered in battle. All of them died on post from illnesses including cholera, rheumatism and typhoid fever. Of those who enlisted in 1867, twenty signed on for a second enlistment. Eight would serve at least thirty years, retiring at the end of the nineteenth century. Those entering their second enlistment could be considered "old soldiers," and they would form the core of Company H in the coming years of conflict against the Indians in Texas, Oklahoma, New Mexico, and Arizona.

BUFFALO SOLDIERS IN THE RED RIVER WAR 1872-1875

Colonel Benjamin H. Grierson
Courtesy Fort Concho Library and Archives.

On July 18, 1872, Company H transferred to Fort Gibson in the Indian Territory as part of Colonel Grierson's plan to re-establish that post as headquarters of the Tenth Cavalry. Far away in northern California, the army battled the Modoc tribe who had been tactlessly placed on a reservation beside their traditional enemies, the Klamaths. Yet in the Indian Territory, no major outbreak took place that year. That did not mean that the Indians had accepted life on the reservations. Official reports stated, "During the year 1872, no general Indian war took place in the division, but the number of murders and depredations committed by small war parties in various places was greater than during the preceding year." Colonel Anderson Nelson claimed, "I have taken measures to learn the indications among the Arapahoes, Cheyennes and Kiowa tribes of Indians, and am well assured, that hostilities are not impending, and that the indications of a friendly and peaceful relation are good." Despite such indications, Company H remained in the saddle through much of 1872. On September 18, Lieutenant Louis Orleman set out with a detachment from Fort Gibson to scout along the Kansas border. Carpenter joined him with the rest of Company H in October. They established a camp near Parker, Kansas from which they could "patrol the northern line of the Indian Territory and remove intruders."[59]

Carpenter's men returned to Fort Gibson on November 30 without seeing any action. On December 23, twenty-eight of the company horses caught influenza. An epidemic of equine influenza had begun in Ontario, Canada in September and would spread across North America by the spring of 1873. Up to seventy-five percent of the horses in the United States became ill. Symptoms included sneezing, hacking cough, depression and loss of appetite. The soldiers of Company H understood that a strong, healthy horse could actually mean the difference between life and death in the field. Twice a day on every post, they responded to Stable Call to groom and feed their mounts. Even when not on patrol, they were continually in the saddle for drill and parade. With their horses being such a central part of military life, the influenza epidemic proved disastrous to the efficiency of the unit. The weary troopers spent their Christmas season caring for the sick horses and bringing them back to full strength.[60]

On April 12, 1873, Company H received orders to return to Fort Sill. Along the way, on April 30, Lieutenant William Harmon and a detachment of eleven men attacked a band of Mexican thieves, recapturing thirty-six horses. Harmon had been detailed to apprehend thieves on several other occasions and had even been cited in orders for his "daring gallantry, energy, and perseverance." After arriving at Fort Sill a battalion consisting of Companies B, G, H and I took the field under Captain Carpenter scouting 333 miles along the line of the Pease River during the month of July. Just after their return, on July 26, John Claggett deserted the troop while under suspicion of petty larceny. Claggett was on his second enlistment and would probably have not deserted except for his anxiety over the charge of theft. To emphasize the importance of military discipline, the regiment offered a thirty-dollar reward and Claggett was apprehended at Fort Gibson on October 7. He pled guilty, received a five-year penitentiary sentence and a dishonorable discharge. While Claggett was being sought, a tragedy occurred in Company H when Private George Hunter was accidentally shot on post. He died in the hospital on August 7, 1873. The loss of such experienced soldiers always proved to be a setback for the company since new recruits needed extensive training and field experience before they became effective soldiers. Still, during the mid-1870s, Company H actually benefited from an increase in enlistment.

Carpenter had already received fifteen new recruits in February. The Panic of 1873 caused by a cascading array of economic disasters including over-speculation in the railroads and the demonetization of silver eventually led to over fourteen percent unemployment. As had been true in most periods of economic hardship, this led to increased rates of military enlistment that lasted through the remainder of the 1870s.[61]

Forrestine "Birdie" Cooper
Courtesy Fort Concho Library and Archives.

Even during the periods of hardship and conflict, there were some moments of peace and prosperity. In writing about her experiences during 1873, Lieutenant Cooper's daughter Birdie later recalled an occasion when the officers of the Tenth Cavalry invited the reservation Indians to a feast on the parade ground at Fort Sill. The Indians entered the post singing and dancing while soldiers "ripped the tops from boxes and barrels" revealing many delicacies that the officers had purchased for the festivities. Cooper thought, "It seemed particularly strange that at any hour the same officers might be ordered to follow and capture these Indians 'dead or alive.'" That actually did prove to be the task facing Company H as they entered the last half of 1873. The troopers joined a battalion under Lieutenant Colonel John Davidson on August 19 scouting the Red River and returning to Fort Sill by the north side of the Wichita mountains on September 14. At the post they found even more recruits to replenish ranks depleted by expired enlistments and desertion. Some of the new soldiers proved ill equipped for army life and could not cope with the reality of frontier service. Three of them, John Dillard, Henry Osborne and David Williams deserted on October 5 without ever having been on patrol. The other recruits began the year 1874 in the field, marching with Company H to temporary quarters at Camp Auger on the Red River. Camp Auger served as their field headquarters through much of the spring and summer. The troopers patrolled against renegades, provided escort for officers and supplies, and enrolled Indians coming onto the reservation. In a rare moment on post, First Sergeant George Goldsby married Ellen Beck at Fort Sill on July 4. The Goldsbys were to have two sons, Crawford and Clarence, and a daughter, Georgia.[62]

Even while Goldsby celebrated his marriage, the first blows in a long expected Indian uprising had been struck. Comanches, Kiowas, Cheyennes, Arapahoes and all the tribes who depended on the buffalo for food, clothing and tools had become increasingly concerned as buffalo hunters systematically moved from herd to herd slaughtering the beasts for their hides and leaving the carcasses to rot on the plains. Between 1872 and 1874 at least four million buffalo were killed. Hunter J. Wright Mooar later insisted, "Buffalo hunting was a business and not a sport. It required capital, management and work, lots of hard work, more work than anything else. Many magazine and newspaper articles claim the killing of the buffalo a national calamity and accomplished by vandals. I resent their ignorance." With two such widely different viewpoints as those of Indians and hunters, conflict seemed inevitable. That conflict reached a head when buffalo hunters including Mooar followed the herds south into the Panhandle of Texas, land that had been granted to the Plains Indians in the Medicine Lodge Treaty. Inspired by Comanche medicine man Isa-tai, who believed himself invulnerable, and led by the half breed Quahada war chief, Quanah Parker, over 600 Comanches and Cheyennes attacked the Adobe Walls trading post in the Texas Panhandle on June 27, 1874. The saloon keeper at Adobe Walls, James Hanrahan had been warned of the coming attack and the Indians failed to catch hunters at the trading post unaware. Most of the buffalo hunters were awake repairing a broken ridge pole when the Indians attempted to surprise them with a dawn attack. Twenty-eight hunters held the warriors off for several days. By the fifth day their ranks had grown to almost one hundred and the Indians were forced abandon their raid. Many of the Comanches and Cheyennes became disheartened when hunter, Billy Dixon, shot an Indian off his horse from almost a mile away. Yet the Indian warriors were not ready to admit defeat. They continued to strike smaller camps of hunters in the Panhandle area. About the same time, Lone Wolf's Kiowas attacked a company of Texas Rangers, partially in retaliation for the death of his son and nephew at the hands of the army earlier that year.[63]

Up until that time, soldiers had not been allowed to pursue renegades who took refuge on the reservation without a request from the Indian agent and they could not operate outside their assigned department. In an attempt to end the increasing number of raids on the South Plains, General Sherman requested permission to pursue the Comanches and their allies wherever they went, even onto reservation lands. On July 20, the Secretary of War replied by ordering the punishment of all those considered renegades without regard to reservation boundaries. The Commissioner of Indian Affairs, Edward Parmelee Smith replied, "But it is a serious problem how to punish the guilty ones without striking the innocent. It is also certain, that, on the opening of hostilities, a large portion of the tribe would leave the agency and take to the plains, when the difficulty of reaching and controlling them by military force becomes greatly increased. It is believed, however, that there is no alternative. The reservation cannot be made a refuge of thieves and murderers." The difference between friendly and hostile Indians proved to be an ongoing problem at Fort Sill. In July, Colonel Davidson issued orders demanding that peaceful tribes "must form their camps on the east side of Cache Creek at points selected by the Agent" and that "No Indian hereafter will be permitted to approach this post nearer than the Agency..." He continued "When friendly Indians desire to visit the Post Commander they must come from the direction of the Agency, and with a messenger from the Agent stating the Chief and the number of his party."[64]

The Tenth Cavalry set August 8 as the date for peaceful Indians to come in to the reservation at Fort Sill and be enrolled. After the deadline already had passed Comanche chiefs Isa-nanica, Little Crow, Black Duck and Big Red Meat asked that their villages be enrolled. Colonel Davidson agreed to admit only Isa-nanica, believing the others had been involved in the Adobe Walls fight. Big Red Meat then moved his sixty Nokona lodges to the Wichita Agency at Anadarko and insisted that they provide him with rations. With only one company of the Eleventh Infantry under his command at Anadarko, Captain Gaines Lawson called upon Davidson for assistance. Lawson, a Medal of Honor recipient, clearly found his company overwhelmed by the sudden appearance of so many potentially hostile Comanches demanding food and supplies. Marching through the night, Davidson arrived at the Wichita agency on the day rations were to be issued, Saturday, August 22 with Companies C, E, H and L and Company I, Twenty-fifth Infantry.[65]

Satanta's people at Fort Sill in 1872. Courtesy of the Oklahoma Historical Society.

Colonel Davidson demanded Big Red Meat surrender and allow his men to be disarmed and returned to Fort Sill as prisoners of war. The chief initially agreed, but a further argument arose when the Comanche warriors refused to surrender their bows and arrows. Meanwhile the Kiowas under Lone Wolf taunted Big Red Meat where he sat on his horse with his bridle held by a soldier, claiming they would fight the troopers if he would fight. Without warning, the chief somersaulted from his horse and ran into the crowd of Indians. Soldiers shot after him and the Comanches and Kiowas returned fire. Davidson formed his men to attack the Nokona village, but the Kiowas firing from behind the soldiers confused them and some of the troopers fired into the friendly Penateka Indians by mistake. Davidson then proceeded to

48

dismount his men at the Washita River and advance against the Kiowas. Some of the Indians fled to the cliffs, while others surrounded the post trader's store situated on a nearby bluff. Seeing the Kiowas looting the store and beginning to retreat with all the supplies they could carry, Davidson recalled, "I ordered up three companies (Tenth Cavalry), under Carpenter, who advanced rapidly up the western slope and drove off the Indians, variously estimated at from 200 to 300, who had already reached the plateau."[66]

Carpenter's men took cover behind boulders and fired by volley. Afraid that more soldiers were on the way, the Indians took their loot and fled. Carpenter remounted Company H and rode through their center scattering a large force of warriors attempting to take a stand behind the troops. Some of the fleeing Indians stopped to kill four civilians gathering hay for forage on the surrounding prairie and lit grassfires to cover their escape. The troopers set counter fires and managed to control the burning hay before flames could spread to any of the buildings. Captain Carpenter reported leading his men in action against about 400 Comanches and Kiowas, "defeating them and driving them from the agency with considerable losses." Four soldiers received wounds in the battle. In Company H, Sergeant Louis Mack was wounded in the foot and one horse was injured. Following the battle, Colonel Davidson cited Mack, First Sergeant George Garnett, Trumpeter Jackson Ferrer and Farrier Pollard Cole of Carpenter's company for gallantry in action.[67]

On August 28, Davidson's battalion returned to Fort Sill to find the army in the process of preparation for the coming campaign. General Sheridan planned to place five columns in the field converging on the canyons at the edge of the staked plains from all directions. Colonel Nelson Miles would march south from Fort Dodge with eight companies of the Sixth Cavalry, four of the Fifth Infantry and scouts, including Delaware Indian trackers; Major William Price from Fort Bascom led four companies of the Eighth Cavalry east along the Canadian River; Lieutenant Colonel Davidson moved west from Fort Sill with six companies of the Tenth Cavalry, three of the Eleventh Infantry, a battery of mountain howitzers and Lieutenant Richard Pratt's Tonkawa guides; Colonel Ranald Mackenzie left Fort Concho with eight companies of the

Fourth Cavalry, four companies of the Tenth Infantry, one of the Eleventh Infantry and the Seminole-Negro Scouts marching north to his old supply camp on the Freshwater Fork of the Brazos River; and Lieutenant Colonel George Buell marched from Fort Griffin with four companies of the Ninth Cavalry, two of the Tenth, and two of the Eleventh Infantry moving northwest to operate between Mackenzie and Davidson with his camp near where Wanderer's Creek empties into the Red River. On September 10, Davidson's column left Fort Sill after two weeks preparation marching along the Washita and then toward their old camp on Otter Creek. The Indian campaign known as the Red River War had begun.[68]

Success depended on the swift movement of large columns of troops and equipment on a scale that had not been attempted in previous campaigns against the Plains Indians. Each column needed to make effective use of both cavalry and infantry as well as supplies for a lengthy period under harsh conditions in the field. The summer of 1874 had been exceedingly hot with many days in August reaching over 100 degrees. Many of the waterholes had dried up and grass for forage had withered. Supplying such large columns would prove to be a significant challenge. Colonel Miles' command made the first contact with the renegades at the edge of the staked plains on August 30 after a quick pursuit in which he abandoned his supply wagons and pressed forward with only what the troopers could carry. In a five-hour fight, he pushed the Indians back over twelve miles, at one point sending Company I to charge with pistols only and engage the enemy at close quarters. The southern column under Colonel Mackenzie had left Fort Concho on August 23 marching north to the previously established supply camp of the Fourth Cavalry on the Freshwater Fork of the Brazos River. On September 19, they left the camp to follow recent trails located by the Seminole-Negro Scouts. Comanches attacked the column on the night of September 26 in Tule Canyon. Aware of the Indian presence in the area, Mackenzie had ordered his horses staked out to prevent a stampede while the men slept with their clothes on. Seeing that they could not panic the army's horses, the Indians withdrew, but continued to fire on the camp from the darkness. Just after dawn, Company E, Fourth Cavalry mounted and charged the attackers, pursuing them for about three miles. The following night, Mackenzie waited until after dark and then marched his men northwest until they found a clear trail leading to the edge of the canyons of the Red River. Seven hundred feet below, Comanche, Kiowa and Cheyenne villages spread out along the floor of Palo Duro Canyon.[69]

On September 28, Mackenzie's troops worked their way down the steep escarpment and attacked the villages in Palo Duro Canyon. Instead of pursuing the Indians when they fled the encampments, he ordered piles of supplies burned and hundreds of captured lodges destroyed. His men rounded up the warriors' horses and slaughtered over a thousand of them to prevent their recapture by the Indians. After several years of campaign experience, Mackenzie had learned that the most effective way to defeat the renegades was to destroy their resources so that they would have nowhere to go except the reservation. In early October, Satanta and Big Tree who had been released from prison the year before, surrendered a band of 145 Kiowas at the Cheyenne agency. The once boastful Satanta lay down his arms, saying, "I am tired of fighting and do not want to fight any more."[70]

Satanta was not the only war chief to return to the reservation. In the aftermath of the fight at Palo Duro Canyon, Colonel Buell's column, resupplied and prepared for a lengthy campaign, had followed the trail of Kiowa and Cheyenne bands along the Salt Fork of the Red River finding at least three abandoned villages and destroying piles of supplies. In each case, Buell discovered trails leaving the camps and continuing in the direction of the reservations. On October 24, Major George W. Schofield, designer of the Schofield Smith & Wesson revolver, attacked a Comanche village at Elk Creek with three troops of the Tenth Cavalry. The surprised Comanches surrendered with several chiefs including Big Red Food, sixty-nine warriors, 250 women and children, and over 1500 horses. That same day, Captain Carpenter with Companies H and L discovered the trail of about fifty Kiowas herding 200 head of livestock near the source of Pond Creek. The troopers immediately took up pursuit causing some of the raiders to scatter and they continued to press along their trail "so rapidly that most of the Indians were compelled to surrender at Fort Sill." On October 26, an additional twenty-five Kiowa warriors with women, children and fifty horses surrendered to Carpenter's command.

After resting at Fort Sill for two days, Companies H and L set out to rejoin their command arriving at Elm Fork on October 31.[71]

November 7 found Davidson's command at the head of McClellan Creek on the trail of a large party of renegades who had attacked a company of the Eighth Cavalry the previous day. Lieutenant Frank D. Baldwin with men from D Companies of the Tenth and Sixth Cavalry located a Cheyenne village on November 8 on the North Fork of Red River. In the ensuing attack, a village of seventy-five lodges was destroyed and two young captives, sisters Julia and Adelaide German were rescued when they were found hiding under a pile of buffalo hides. Most of the Cheyennes escaped to the west. Baldwin later received the Medal of Honor for routing a superior force from their strong defensive position. Mounting Companies B, C, F and H and with detachments of Companies E and I, Eleventh Infantry and Lieutenant Pratt's scouts, Captain Charles Viele followed the Cheyennes "for a distance of ninety-six miles, having several slight skirmishes with the rear guard of Indians and capturing a number of ponies and mules, the latter packed, which the Indians had abandoned in the flight." On the fifth day of pursuit "began a violent rainstorm, changing to sleet and snow, which lasted until November 19, causing the death of one hundred horses from insufficient food and cold, and freezing the feet of twenty-six men." John Casey later first sergeant of Company H recalled, "During this trip each man was only allowed to carry one blanket and his poncho and we had to drive our picket pins between the pommel and canta (cantle) of the saddle. In this way we had to sleep with our heads upon our saddles and the horses kept us awake and uncovered most of the night. It was very cold and we nearly froze."[72]

When the "Norther" hit, Captain Viele's battalion turned back toward camp, unable to follow the Cheyenne trail any farther. Casey remembered the return trip: "It rained and snowed continually all day and night and we had no overcoats, only our ponchos. When we reached headquarters, about 12 o'clock at night, the snow was six or eight inches deep. In this we had to lay down to sleep with our clothes and blankets wringing wet and the weather continued very cold during the entire time we were there." He continued, "In this campaign the regiment was for ten days snowed in without rations but one hardtack a meal and very little coffee because our supply train was lost in the snow storm." The frozen troopers tried to start fires using wet buffalo chips. When that failed, they cut down cottonwood trees and built blazing bonfires from the trunks while the horses ate the branches. On November 29, Davidson's column finally limped back in to Fort Sill. Aaron Archer and Andy Clayton of Company H both received treatment for frostbitten toes. Benjamin Bard had frostbite so severe that Captain Carpenter allowed him to ride in a supply wagon on the way back to the post. Before he could fully recover, Bard was assigned guard duty on a "cold sleety night with rain freezing as fast as it fell." With his feet now hopelessly frostbitten, Benjamin Bard spent most of the remainder of his enlistment in the hospital and was discharged with a fifty percent disability. Despite severe weather conditions, Davidson's men had accomplished their purpose. They had destroyed several camps and captured nearly 400 renegade Indians and over two thousand horses and mules without losing a single soldier.[73]

Skirmishes with renegade bands continued throughout the rest of that difficult winter. Companies of the Tenth returned to the field in December in spite of bitter cold conditions. On December 18, Captain Alexander Keyes with Company D encountered a band of Cheyennes on Kingfisher Creek and captured fifty-two men, women and children and fifty animals. In January 1875, the temperature reached twenty-five degrees below zero. Some of the Panhandle streams froze so hard that army supply wagons could cross the ice fully loaded. On February 23, 1875, Colonel Davidson's men attacked a Kiowa village on the Salt Fork of the Red River. They captured sixty-five warriors, 175 women and children, 300 ponies and seventy mules. Among the prisoners were several prominent chiefs including the once defiant Lone Wolf. In March, the major body of the Cheyennes under Stone Calf came to the reservation to surrender. Remnants of the Comanche, Kiowa and Southern Cheyenne nations gradually drifted back to the reservation lands around Fort Sill and Medicine Bluff Creek during the spring of 1875. Many Cheyenne and Comanche warriors might have continued to resist, but most of their women and children were sick and starving. Captain Carpenter claimed the long campaign had "discouraged the Indians, and they surrendered, agreeing to give up horses and arms." Knowing that some might attempt to leave the reservation after replenishing food and supplies, the women and children were kept under close watch while many of the warriors were confined to the guardhouse. Some of the leaders like Lone Wolf were placed in irons and shipped to the prison at Fort Marion in Saint Augustine, Florida. The army returned Satanta to the penitentiary at Huntsville, Texas where he took his own life in 1878 by leaping headfirst from a window of the prison hospital. Big Red Meat had already sickened in captivity and died in his sleep on January 1, 1875.[74]

While some of the warriors were being prepared for transport to eastern prisons, another outbreak occurred. On April 6, a group of Cheyenne prisoners were taken to the blacksmith to be fitted with leg irons while some of the Indian women stood nearby mocking them. Suddenly, Black Horse kicked the blacksmith and raced away toward the Cheyenne camp, not slowing even when the guards called for him to halt. The soldiers fired, killing Black Horse and hitting several of the Indian lodges. Over one hundred Indians fled toward the nearby sand hills where they dug up arms and ammunition they had hidden there and established a position. The soldiers proved unable to dislodge them and a Gatling gun was brought up to spray fire on the Cheyennes. That was to have been followed by a charge on the hills, but Captain Keyes to the southwest of the Indian position could not find any terrain upon which to advance on horseback. He did attempt a second advance on foot to support Company M, Sixth Cavalry, but that stalled under heavy fire. Lieutenant Colonel Thomas Neill, who had ordered the advances, later reported that the companies of the Tenth Cavalry had refused to fight. Although the evidence did not support that claim, Neill's report did damage the reputation of Keyes' command. The attack was halted and ordered to resume the following morning, but during the night, the Cheyennes slipped away heading north into Kansas in an attempt to join relatives in Montana. Troopers pursued the Indians almost four hundred miles without overtaking them. Lieutenant Austin Henely with Company H of the Sixth Cavalry took up the chase from Fort Wallace. On April 23, he finally came upon one of the Indian bands resting their exhausted horses at a site along Sappa Creek in northwestern Kansas. Henely attacked the camp, killing twenty-seven Cheyennes including women and children. This encounter at Sappa

Creek proved to be the final action of the Red River War. In June 1875, the last band of renegades from the reservation, a group of Quahada Comanches led by Chief Quanah Parker surrendered at Fort Sill.[75]

By the time Quanah surrendered, the Tenth Cavalry already had left Fort Sill. With the Red River War coming to an end, many of the regiments had been reassigned. Colonel Mackenzie and the Fourth Cavalry took over the posts guarding the reservations in Indian Territory while the Tenth was on its way to the west Texas frontier. On a plateau overlooking Sweetwater Creek in the Texas Panhandle where only a few months before Colonel Miles had established an advance supply camp, a permanent post was being built. That post would become known as Fort Elliott and would bring settlers to the area beginning with the founding of nearby Mobeetie. The fort guarded the Panhandle area until 1890 by which time new settlements had begun to grow at the sites of the future cities of Lubbock and Amarillo. Only one year after the Red River War, Charles Goodnight drove the first cattle into the Panhandle. With the buffalo gone, many cattlemen would follow taking over the wide-open range. By the 1880 census, twenty-six counties had been formed with a population of 1607. Only fifty-one Black men lived in the Texas Panhandle five years after the Red River War. Of those, thirty-six of them were stationed at Fort Elliot.[76]

When the Tenth Cavalry left Fort Sill in 1875, they already were exhausted from a bitter winter in the field and lacked adequate men, horses and equipment to embark upon their new assignment of building roads and telegraph lines, guarding supply trains and patrolling the vast unsettled lands of western Texas. Although they continued to receive recruits regularly, the mounted companies never reached full strength during their tour of duty in Texas. The Tenth Cavalry headquarters and band left Fort Sill along with Companies B, C, H, I and K for their respective stations in Texas on March 27. The remaining companies would follow by summer. As they left Indian Territory, Company H rode away from the rest of the column swinging further south and west to Fort Davis in the Trans Pecos region. That lonely outpost in the west Texas mountains would be their home for the next ten years.[77]

TRAVELING AN UNKNOWN TRAIL:
FORT DAVIS, 1875-1880

Captain Louis Henry Carpenter

In February 1875, Congress passed a civil rights act guaranteeing Blacks equal treatment in public accommodations and transportation and equal opportunity for jury service. Although an important step forward, this law probably had little effect on the Black troops along the western frontier. It was not enforced in most cases and sections of the act were struck down by the Supreme Court in 1883. It definitely did not improve conditions for the Indian tribes against whom they fought, even with the end of the Red River War. In the northern and western territories, the battle to bind all Native American tribes to reservation lands continued. When the Sioux and Northern Cheyennes refused to leave their winter quarters in December 1875 to settle on a reservation, General Sheridan mounted an offensive against them. Once again, he used a multi-column attack. One column under General George Crook fought over 1000 Indians at Rosebud Creek on June 17, 1876. Colonel George Custer with the Seventh Cavalry advanced into the same area the following week, unaware of Crook's encounter. Custer decided to engage the Sioux in the valley of the Little Bighorn without waiting for a cooperative effort by the other columns. On June 25, a detachment of 230 men including Custer was surrounded and killed by over 3000 warriors. In the Arizona and New Mexico territories, Mescalero, Chiricahua and other Apache tribes were forced to live together on reservations without regard to their traditional homelands or differing ways of life. Beginning in 1875, when they left those reservations to carry out raids across western Texas and into Mexico, they became the business of the newly transferred Tenth Cavalry.

On May 1, 1875, Company H broke camp beside the waters of Limpia Creek and rode the final four miles through the mountain passes to Fort Davis, Texas. Riding along the "always clear, pure, and cool" creek bordered by cottonwoods and willow with a rich profusion of wildflowers scattered about, the soldiers had their first opportunity to survey the vast spaces of western Texas. Fort Davis, located at a strategic point along the San Antonio - El Paso stage road, had been an important link in the frontier line of defense since 1854. A small collection of buildings, many constructed of adobe plastered inside and out, settled against a backdrop of soaring cliffs in the canyon of the Limpia. This remote outpost would be home to Company H for the next ten years. Captain Carpenter led the small column of fifty-two troopers, which included soldiers who had been with the company since 1867, like Trumpeter Silas Jones and Sergeant Pollard Cole. Over the past eight years, they had patrolled and fought in every corner of Kansas and the Indian Territory. At Fort Davis, they would continue to build their reputation as one of the most effective and dedicated companies of cavalry to serve on the western frontier.[78]

Captain Carpenter's company soon became even more familiar with the West Texas landscape. Although they quartered at Fort Davis for ten years, the troopers only spent about half of that time on post. John Casey, later company first sergeant claimed, "I was on continuous scouting from ten to thirty days, and from that to a year and a half at a period." On one of their typical patrols in 1875, Company H reported thirty-seven enlisted men and two officers present for duty, eleven men on detached service, six on sick call and one under arrest. Captain Carpenter spent part of that first summer at Fort Lyon, Colorado testifying as a court-martial witness. He was delayed returning to his troop by "serious washouts on the railroads" which forced him to take a roundabout route to Fort Davis going through Saint Louis and down the Mississippi River to New Orleans. Second Lieutenant Charles G. Ayres was assigned to Company H in September 1875. Joining the army in New York, he had been commissioned second lieutenant in the Twenty-Fifth Infantry on October 31, 1874. In evaluating Ayres, Captain Carpenter commented, "He was always faithful in the performance of his duties and perfectly subordinate. Very loyal in carrying out the wishes and intentions of his troop commander." He performed "very active and arduous duty against Indians in North Western Texas."[79]

With the Kiowas and Comanches finally settled on reservations, the Tenth Cavalry now faced a new challenge. Apaches, mostly Mescalero, roamed freely across western Texas from the Guadalupe Mountains deep into the Mexican state of Chihuahua. On September 3, 1875, Captain Carpenter departed the post leading his troop along with Company I, Twenty-fifth Infantry into the Sacramento Mountains of New Mexico to scout for Apache raiding parties. They returned to Fort Davis on October 21, having traveled 679 miles, climbing through Guadalupe Pass and exploring many of the trails between Eagle Springs and Fort Quitman to find a shorter, more effective route for future operations. The following month, they camped along the Rio Grande patrolling as far as Presidio. While there Carpenter received information that the Mexican government had made an agreement with Mescalero chief Alsate allowing him to trade at Presidio and camp in Mexico, raiding the American side and "carrying stock and other spoils to Mexico to trade with impunity at San Carlas and Presidio del Norte." In February 1876, Company H struck a three-day old trail running toward the Guadalupes. They followed a large party of Indians with approximately 300 head of stock "through difficult canyons and over rocks and high mountains." On the second day, the trail divided, with one branch leading toward the Sacramento Mountains and the Mescalero reservation. Carpenter believed this was further evidence of Alsate's raids. He proposed a cooperative effort between the cavalry and Colonel Joachim Terrazas' Chihuahuan state troops against the Apache chieftain, but no action appears to have been taken on that proposal. Major George L. Andrews did march two companies of the Twenty-Fifth Infantry to Presidio del Norte in December taking with him a three-inch cannon and a detachment from Company H including Joseph Claggett. In an effort to protect American merchants and secure the release of a U.S. citizen taken hostage south of the border, Major Andrews fired his cannon into the town of Presidio. Claggett claimed that he acted as "lanyard-puller" and that the cannon produced "such excellent results" that fifteen Mexicans surrendered.[80]

During 1876, Company H had detachments on patrol for eight months out of the year. Accompanied by Companies H and K, Twenty-fifth Infantry, the company trailed Apaches through the Carrizo and Guadalupe Mountains returning to post March 30. Successive detachments led by Lieutenant Ayres, First Sergeant George Garnett, and Sergeant James Campbell followed the elusive track of Indian raiders and stolen stock throughout the summer. On June 27, Captain Carpenter took leave to visit the great Centennial Exposition in his home town of Philadelphia. First Lieutenant William Harmon had fallen from a rearing horse in May 1876 and fractured his skull leaving Lieutenant Ayres in command. After being placed on sick leave, Harmon's condition continued to deteriorate and was further complicated by tuberculosis. Harmon never rejoined the company. He eventually received a disability retirement and died on December 12, 1886, in Cincinnati.[81]

Despite constant patrols monitoring all known sources of water, Apache raids continued throughout the Trans-Pecos region of Texas. On October 9, 1876 stage employee Juan Marengo was killed at the Eagle Springs mail station. On March 7, 1877 two men were found dead only four miles from Fort Davis. In June 1877, Lieutenant Ayres' command discovered a recent Apache camp in the Guadalupe Mountains. They pursued the elusive Apaches through the mountains and recaptured thirteen head of stolen stock. During that same period, John Casey remembered riding into Musquiz Canyon to relieve a party trapped by Indians. In the early morning darkness, Company H charged through the narrow canyon in "columns of fours." Casey's horse stumbled in a hole and two other horses fell over him, dislocating Corporal Casey's shoulder. Lieutenant Ayres later recalled in a letter to Casey the shock of "you falling into the well with three or four other men and horses on top of you, and the wonder that you were not killed." Enduring constant pain, John Casey stayed on the raiders' trail for an additional six days before he could return to post.[82]

Casey's injury while actually in pursuit of Indians proved to be unusual. Far more often, troopers reported to sick call after being bitten by a horse or cut and beaten in a barroom fight. In 1879, the post surgeon treated Charles Faulkner for a sprain resulting from "falling over a tent pin" in the dark. H Troop's soldier-preacher William Allen had a revolver fall from his pocket, strike a table and discharge, wounding him in the right side while off duty. The most common illnesses at Fort Davis seem to have been diarrhea, dysentery and rheumatism. The standard army ration of bacon, hardtack, beans and coffee brought many more soldiers to the post hospital than wounds suffered in Indian attacks.[83]

On August 2, 1877 the body of stage driver Hank Dill was brought in to Fort Davis by the guard at the El Muerto Springs stage station located about forty miles west of the fort. Dill had been killed by Indians the day before while herding mules near the station also known as Deadman's Hole. Sergeant Joseph Claggett, a veteran of ten years with Company H, assembled ten men and a scout to follow the raiders' trail from El Muerto. The trail remained clear until they reached a point about five miles north of Van Horn wells. There the herd scattered. The only source of water Claggett knew within a reasonable distance was Rattlesnake Springs, so he led his men north along the Sierra Diablo range. The spring proved to be dry that year, so the patrol turned northeast riding into the Guadalupe Mountains. Along the trail, they did find an old canteen and signs of beef being slaughtered which led their guide to believe the raiders were New Mexico reservation Indians. After 48 hours without water or any further sign of the raiders, Sergeant Claggett's command finally turned back toward Fort Davis.[84]

During this period Captain Carpenter commanded six troops of the Tenth encamped on Pinto Creek near the Rio Grande for a possible campaign into Mexico. Once again, no cooperative operation developed between the two countries. After drilling and preparing the troops for several months, he returned to post on October 23, 1877. On December 17, he marched Company H to San Elizario near El Paso "to assist forces in quelling riotous and turbulent conduct of Mexicans residing there and in the vicinity." The El Paso Salt War had resulted in riots and several executions over whether fees should be paid to El Paso businessmen for salt gathered from the salt flats ninety miles east of the city. Many of the local citizens on both sides of the river had gathered salt for years without paying a fee and resented what they believed was a corrupt maneuver to gain title to the flats for the benefit of a few officials. On October 10, 1877, Judge Charles Howard walked into an El Paso store with a double-barreled shotgun and killed Louis Cardis, a former associate in the salt scheme that he had quarreled with on several occasions. Howard then proceeded to file suit against sixteen wagons he knew were headed for the salt flat during the first week of December. He also called upon a newly recruited company of Texas Rangers under Lieutenant John Tays to enforce the ban on salt unless all fees were paid. Angry citizens trapped Howard and the ranger company in San Elizario and kept them under siege for five days. Howard surrendered his men on December 17, the same day Troop H rode out of Fort Davis to provide them with assistance. The mob executed Judge Howard and two associates. They released the rangers without arms. This proved to be the only time an entire company of Texas rangers had been forced to surrender. Troop H did not arrive until December 23 after most of San Elizario had been looted.[85]

Major George Andrews commanding Fort Davis wrote to departmental headquarters, "The organized rioters were over three hundred and fifty strong, accompanied by about one hundred and fifty thieves and vagabonds, these last have done a great deal of stealing..... The Mexican Authorities posted a large number of notices on both sides of the river, warning Mexican Citizens against taking part in the trouble." But, he continued, "Most of the rioters have escaped across the Rio Grande." Some indictments were brought against the leaders of the mob, but there were no arrests. Company H maintained an uneasy peace patrolling the streets of San Elizario and nearby Isleta. On January 18 they left and returned to Fort Davis. Possibly the most significant outcome of the Salt War incident was the reactivation of Fort Bliss in El Paso.[86]

While Company H guarded against looters, the Apaches continued to raid the Trans-Pecos area. On December 23, 1877, stage agent Gabriel Valdez and Horan Parsons were attacked and killed in Bass Canyon while hauling water from the Eagle Springs station to Van Horn wells. Then on January 5, 1878, Mescaleros killed six men in a raid on a ranch along the Rio Grande about sixty miles northwest of Presidio del Norte. Between February and April at least five men died at Point of Rocks in Limpia Canyon, only eighteen miles northeast of Fort Davis. To address this growing concern, the Department of Texas created the District of the Pecos in 1878 with Colonel Benjamin H. Grierson of the Tenth Cavalry as district commander. Grierson's strategy for controlling western Texas required that his troops not only scout against the Indians, but also map unexplored regions and identify sources of water and forage. To deal with the Apache problem in the Trans-Pecos region, he decided to establish a line of subposts at strategic locations throughout the district. In March, he dispatched Company H to locate a permanent post "at some point on the Eastern slope of the Guadalupe Mountains" where water and grass were always plentiful. Carpenter had mapped the region on previous scouts and was familiar with the area at the summit of Guadalupe Pass where many good springs could be found. As the initial site for an outpost, he selected Bull Springs near the old Pine Springs overland mail station.[87]

Pine Springs mail station in 1858. Drawing courtesy Guadalupe Mountains National Park.

Colonel Grierson continued to emphasize the importance of eliminating the Apaches' ability to resist, stating, "Troops sent out in pursuit of Indians must be amply provided for a long & vigorous pursuit. The Indian marauders must be attacked wherever found and severely punished if possible." On May 20, the troopers of Company H left Fort Davis again to scout the Carrizo Mountains and establish a camp in that area. Taking five pack mules loaded with scouting rations for a month, "half rations of short forage" for the animals and 100 rounds of ammunition per man, the troopers traveled along the road or by night to avoid contact with the Apaches. Determining that there was "no reliable water suitable between the Rio Grande and the Guadalupe Mountains, excepting Eagle Springs," the water supply used by stage employees at Van Horn Station, Carpenter chose that spring as a supply camp from which to patrol the Sierra Diablo and Carrizo Mountains. On June 6, Carpenter received instructions from District of the Pecos headquarters stating, "The importance of discovering permanent or living water heretofore unknown can hardly be overestimated, and it can only be accomplished if at all by hard work, continued scouting and explorations." He was further advised, "In case, however, you come upon Indians, you will attack them vigorously, at all hazards, and spare neither horses nor men to secure the desired results, which is the destruction of the Indians, or the severest punishment that can be inflicted upon them…" Carpenter's mission became further complicated when he discovered that the Pueblo Indian assigned as his scout had weak eyes and knew nothing of the area. He commented that the troop would have to "trust to fortune in finding water." Good fortune and hard work continued during the long, dry summer of 1878, as the men of Company H explored and mapped numerous springs and waterholes including Alamo Spring, Apache Spring, Rattlesnake Springs and Sulphur Springs. Some of them were intermittent and most rose from faults in the rocky canyons.[88]

Conditions did become increasingly difficult as the arid summer months of field duty continued. On July 24, five troopers, including Sergeant Warren W. Wright, deserted and headed toward the Mexican border. During a period where desertion sometimes reached twenty-five percent, Company H had continued to maintain a remarkably low rate. Carpenter determined that he would not to let the men escape. He ordered a detachment under Lieutenant Ayres to cross the Rio Grande if necessary and pursue the deserters "a reasonable distance" into Mexico. On July 25, Ayres caught them south of the river and captured four of the five soldiers with all of their horses and most of the equipment. The prisoners were escorted back to Fort Davis on August 1 to face court martial charges and dishonorable discharge. Following that incident, not one man deserted from Company H for over two years.[89]

Captain Carpenter received a dispatch from Fort Davis on August 6 informing him that a detachment of Mexican troops had organized an expedition against Apaches suspected of attacking Ruidoso and would be authorized to operate in Texas. Colonel Grierson encouraged full cooperation with this expedition. Carpenter believed the Mexicans would trail the Indians north into the Guadalupes so he determined to follow them by way of Rattlesnake Spring past the salt lakes. On August 13, he found the camp of Captain Norvell and Company M in the Guadalupes. With Norvell was scout Jack Stilwell whom Carpenter and some of the veteran soldiers remembered from the rescue at Beecher Island. Company M was also in search of the Mexican expedition and the Apaches that Stilwell had tracked headed toward for the Mescalero Reservation earlier in the summer. After a fruitless search Company H returned to Eagle Springs on August 17 only to find that the Mexicans had been delayed and had not yet started.[90]

On August 29, Carpenter's troop finally returned to their post after riding 1806 miles on patrol. They had been north as far as the Guadalupe Mountains and south across the Rio Grande. Though ordered to destroy any Apaches they encountered, to "attack them vigorously, at all hazards, and spare neither horses nor men . . ." Company H seldom even sighted an Indian. Carpenter believed that most of the trails they followed could be traced to the Mescalero Reservation at Fort Stanton. He suggested in his scouting report that if the Apaches were removed from the rugged terrain of New Mexico to "the flat regions of the Indian Territory . . . the outrages carried on for such a length of time in North Western Texas would cease."[91]

Yet the flatlands of the midwest did not prevent problems with other tribes. On September 9, 1878, the Northern Cheyenne under Dull Knife left their reservation at Fort Reno. The Fourth Cavalry and Sixteenth Infantry pursued, engaging Dull Knife at Sand Creek, Kansas on September 21-22. The Cheyennes escaped and ambushed soldiers near Fort Wallace on September 27 before being captured near Fort Robinson, Nebraska on October 23. In western Texas, raids on ranches and stage lines increased during 1879 as Victorio and his Warm Springs Apaches escaped the reservation with over 100 warriors and their families. Captain Carpenter commanded the post at Fort Davis from his return in August 1878 through most of 1879. Lieutenant John Bigelow described Carpenter during this period, "I made the acquaintance ... of Capt. Carpenter of the 10th, who is now in command of the post. He is considered the best company commander in the regiment and one of the best in the service. He is a gentleman by birth and training. He is not a narrow minded 'routinier' but a broad minded student of his profession."[92]

Beginning April 1, 1879 Carpenter kept a detachment from Company H on rotating duty at Eagle Springs throughout the summer. Lieutenant Ayres mounted the first guard, stationing twenty men at Eagle Springs and five men at Seven Springs. On July 25, a detachment of twelve men led by Captain Michael L. Courtney of the Twenty-fifth Infantry encountered a raiding party between Sulphur Springs and the Salt Lakes. During the engagement, two soldiers, Corporal William J. Webb and Private George Forster were seriously wounded. Forster received a gunshot to the left breast. Corporal Webb, shot in the right thigh, never fully recovered. Despite these casualties, the troopers continued to drive the Apaches back, killing two warriors and wounding one. They captured ten horses and mules as well as assorted equipment. Joseph Claggett recalled that one of the captured mules remained with H Troop until they were assigned to Montana in 1892.[93]

In September 1879, trouble continued on the northern frontier as well as in western Texas. At White River Indian Reservation, in northern Colorado, the Utes, killed Indian agent Nathan Meeker and nine employees. On September 29, men of the Fifth Cavalry and Fourth Infantry were attacked by over 300 Utes in Red Canyon and trapped within a ring of wagons. Relief from Fort Russell, Wyoming, arrived on October 5 and drove off the attackers. At Fort Davis, scouting patrols continued throughout the fall of 1879. On September 14, the Eagle Springs detachment trailed an Apache band through the Viejo and Van Horn Mountains, actually sighting and shooting at a few Indians along the mountain trails. After turning over command of Fort Davis to Major Napoleon B. McLaughlin on October 23, Captain Carpenter marched his entire troop to Eagle Springs and established camp.

Colonel Grierson had received word that "a large party of Indians were moving southward along the Rio Grande" by way of the spring. Carpenter's orders were "to attack and destroy, or vigorously pursue any Indians who might make their appearance in that sector of the state." Just after their arrival on the evening of October 27, a patrol reported fresh signs had been found using moonlight. Carpenter mounted his men and followed the trail "by the light of the moon, over a very rough country" until almost dawn. It soon became obvious that the signs were several days old and had only appeared fresh under the shadowy moon. Since the company was already on patrol, Carpenter decided to scout all of the trails and watering places that the Apaches might use between Eagle Springs and the Rio Grande. He also consulted with the drivers on the stage road who promised to report any sign they might see on their daily routes.[94]

As Colonel Grierson positioned the Tenth Cavalry for action, the Apaches continued to attack ranches and stage lines throughout the Trans-Pecos region. As leader of the Ojo Caliente or Warm Springs Apaches, Victorio originally had agreed to live on a reservation in his homeland of western New Mexico. Then in 1877, a move toward consolidation of Indian lands forced the Warm Springs band onto the San Carlos reservation in Arizona. San Carlos, in addition to the fact that it was not their home, suffered from extreme heat, sterile soil, disease, and an abundance of reptiles and insects. Conditions proved intolerable, and Victorio escaped on September 2, 1877 taking most of his people with him. They settled on the Mescalero reservation briefly in 1879, but when Victorio learned that he might be arrested for horse stealing and murder, he escaped once more with over 100 warriors and their families. On September 4, his warriors struck the horse herd of Company E, Ninth Cavalry at Ojo Caliente in New Mexico territory. As raids increased, Major Albert P. Morrow with a battalion of the Ninth took the field to capture Victorio. While Morrow chased the raiders throughout the fall, supply problems plagued his command. Perhaps most important, the soldiers became dangerously short of remounts. Supplying sufficient horses in good condition for the Black cavalry had continued to be a problem that was made worse by the Apaches' ability to replenish their herds through frequent raids. By the beginning of 1880,

Victorio's warriors had become experienced at eluding the army and more determined than ever to remain free from the reservation.[95]

On March 23, 1880, Colonel Grierson left Fort Concho with five companies of the Tenth Cavalry and a detachment of the Twenty-Fifth Infantry. The troopers' objective was the Fort Stanton reservation where they were to assist in disarming and dismounting Mescaleros believed to be providing Victorio with supplies. Carpenter remained on duty at Faver's Ranch scouting the line of the Rio Grande with Companies C and H. Grierson's column passed through Pecos Falls, Texas on March 31. There they discovered that Indians had stolen stock from nearby ranches only the night before. Grierson dispatched Lieutenant Calvin Esterly with a patrol to follow the Apache raiders. On April 3, 1880, in the middle of a severe dust storm, Esterly attacked the raiding party and recovered eight head of stock. He rejoined the main column on April 6 near the mouth of the Black River. That same day, Company K under Captain Thomas Lebo set out to scout along the line of march, covering "a belt of country over fifty miles wide." Lebo found extensive signs of Apache raiding parties. He followed one recent trail into the Guadalupe Mountains and on April 9 overtook the Indians at Shake-Hand Springs. During the resulting fight, Company K completely destroyed the Apache camp. They killed the party's leader, capturing four women and one child and recovering a Mexican captive.[96]

Grierson's main column arrived in the vicinity of the Mescalero reservation on April 12, 1880. Heavy storms over the next two days delayed the disarming of Victorio's supporters until the afternoon of April 16. At about 2:30 p.m., Grierson received a signal that the Mescaleros were attempting to escape rather than have their weapons taken. The Tenth Cavalry immediately moved forward across the Tularosa River, attacking the Mescaleros as they fled into the surrounding mountains. The attempt to disarm the Apaches before they could join Victorio had failed. Although Colonel Grierson's troopers captured over 250 Indian men, women and children, many eluded the soldiers. Detachments of the Ninth and Tenth Cavalry pursued the Apaches into the mountains. Lieutenant Mason M. Maxon overtook one group and drove them back toward the reservation, killing one warrior and capturing five horses. Colonel Edward Hatch of the Ninth Cavalry tracked Victorio to near White Sands, New Mexico, but the Indians managed to retreat into Mexico. Colonel Grierson remained in southern New Mexico until April 27, 1880. Then the Tenth Cavalry returned to Texas leaving Captain William Kennedy with Company F to man the outpost in the Guadalupe Mountains.[97]

On May 16, Grierson returned to his headquarters at Fort Concho. He instructed Company M stationed at the head of the North Concho to "keep out detachments in all directions...to give any marauders a warm reception in case they should make their appearance in your section of the country." The camp at Eagle Springs now became the field headquarters for Company H as they received three quartermaster wagons with forage and supplies. According to Captain Carpenter, Companies C and H were "to watch the Indian runways and to have general charge of the line." Colonel Grierson had determined to defeat Victorio, "Old Vic" as he called him, by patrolling all the springs and waterholes in the Trans-Pecos area to keep the Apaches from water and force them across the border into Mexico. Carpenter's position at Eagle Springs, a relatively "reliable" source of water in marching distance of numerous other springs proved vital to that effort. On May 12, 1880, Apaches attacked a settlers' wagon train in Bass Canyon. Company H investigated the scene and followed the Indian trail over Viejo Pass. The raiders had crossed the Rio Grande eight days before, but Carpenter did discover that the Indians had been using a large "never failing spring" located in a rocky gorge in Viejo Pass and he established a camp at that site. On May 19, Carpenter received a telegram from Grierson with orders: "You will proceed immediately with all your available force to Eagle Springs, Texas, from which point you will scout the Cariso and Eagle Mountains and drive all Indians out of the country west of this post and keep them out."[98]

Meanwhile, Colonel Hatch of the Ninth, his troops exhausted by fruitless patrols, requested that the Tenth Cavalry return to New Mexico. Grierson protested and managed to convince General Sheridan that his soldiers would be more effective against Victorio in western Texas. General Edward O. C. Ord backed him in his plan to patrol the springs and waterholes. On June 17, 1880, a battalion of the Tenth Cavalry led by Captain

Nolan arrived at Fort Concho from temporary duty at Fort Sill. Grierson described Nolan's command as "not more than half mounted, poorly armed and equipped, and entirely without pack mules." Despite their condition, the colonel needed Nolan's battalion in the field to keep Victorio away from water. He managed to resupply them in a very short period and by June 27 had issued orders for the command to proceed to the Guadalupe Mountains on patrol.[99]

On July 10, 1880, Colonel Grierson left Fort Concho with an escort which included his teenage son Robert and his adjutant, Lieutenant William H. Beck. They reached Fort Davis a week later and Eagle Springs on July 23. While there, the colonel communicated with Mexican Colonel Adolfo Valle and agreed to intercept any Apaches crossing into Texas. On July 27, Grierson's party arrived at Fort Quitman along the Rio Grande. The next morning, they were shocked to find Colonel Valle's forces on the opposite bank of the river, weary and without provisions. Grierson instructed his quartermaster to issue the Mexican troops 1000 pounds of flour, 500 pounds of oats and 630 pounds of corn.[100]

With Colonel Valle no longer in pursuit of the Apaches, Grierson believed that Victorio would soon cross the river and head north. On July 29, 1880, with an escort of seven men and his son Robert, Grierson set out for Eagle Springs. Near Devil's Ridge, south of present-day Sierra Blanca, they sighted a single Indian. Believing that Victorio might be moving in that direction, Grierson established camp at Tinaja de los Palmas, also known as Eighteen Mile Waterhole. After sending for reinforcements, the small party of soldiers fortified a ridge overlooking the watering place. On a strategic spot, called "Rocky Point" by Robert Grierson, the troopers built two stone breastworks and named them Forts "Beck" and "Grierson."[101]

Around 4:00 a.m. on July 30, Lieutenant Leighton Finley with a detachment of fifteen soldiers arrived to escort Colonel Grierson's party to safety. Instead of leaving, the colonel had Finley build a fortified position lower on the ridge and sent two of his men to speed reinforcements from Eagle Springs. About 9:00 a.m., a force of sixty Apaches appeared. Seeing the cavalry positioned in the rocks, they attempted to cross the nearby stage road and move away from the waterhole. Grierson sent Finley with ten men to draw the Indians into a fight until reinforcements could arrive. Captain Charles Viele with Company C reached the scene about an hour later. His advance party mistook Finley's troops for Indians and fired at them, forcing them back to the rock fortifications. Soon realizing their mistake, Viele's company engaged the Apaches and kept them from moving northward until Captain Nolan's company arrived. With troops attacking them from both sides, Victorio's band retreated and recrossed into Mexico. Grierson's scouts followed as far as the Rio Grande.[102]

Over the dry summer months, Captain Carpenter's command had increased to include Companies B and I and a detachment of the Twenty-fourth Infantry. On June 20, Carpenter reported a shortage of water, grass, wood and fresh meat around Eagle Springs. Because of scant, muddy water and poor rations, a number of the soldiers suffered from dysentery and kidney disorders. Carpenter claimed that "continuous service for cavalry at Eagle Springs is the hardest on men and horses of any duty performed from Fort Davis . . ." To improve conditions, Captain Carpenter moved most of his battalion to the spring in Viejo Pass during July. On July 14, he received a report that Colonel Valle would take the field from Carrizal, Mexico probably forcing Victorio to cross the river. Cavalry pickets on the Ojo Caliente trail encountered about sixty of Victorio's warriors moving north between Eagle Springs and Fort Quitman on July 29. Carpenter mounted his battalion and followed the Apaches along the Devil's Ridge. After riding through a drenching and long-awaited rain, they made camp around 1:00 a.m. The next morning, a courier found them with the news that Colonel Grierson had attacked the Apaches at Tenaja de los Palmas and was on his way to Eagle Springs. Carpenter left Company I under Captain Theodore Baldwin to guard Viejo Pass and turned the rest of his command toward the springs to rendezvous with the district commander. At the same time, he ordered that, "Pickets were thrown out towards the Rio Grande and the country watched closely."[103]

In the dark morning hours of August 2, Carpenter sent one of the picket patrols under Corporal Asa Weaver toward the Rio Grande to locate the Apaches. Weaver, age 27 from Indiana and still in his first enlistment, had already proven his leadership ability. With him went seven men from various companies and several

Indian scouts. One of those men was Private William Brent of H Troop, a former slave from Virginia who had enlisted the previous year. All day on August 2, Weaver's patrol scouted trails leading to Alamo Spring. They established camp at the spring and near dawn the following day, Weaver discovered a large party of Victorio's Apaches advancing toward the water. A fifteen-mile running battle ensued. Every few miles, Weaver would find a good position, dismount his patrol, turn and fire on the Indians. After slowing their progress, he would mount and charge on to the next position. Once while remounting, Private Willie Tockes of Company C lost his reins when an Apache bullet hit his horse. The injured horse panicked, turned and stampeded right into the midst of the Apache warriors. The last sight Weaver had of his trooper, Tockes had his spurs dug in and was firing his carbine at the Indians as his horse blindly charged, reins dangling on the ground. Many months later, another patrol recovered the remains of his skeleton. By the time Weaver's patrol reached camp, every horse had been wounded, one trooper was dead and one had been wounded in the foot Weaver received an immediate field promotion to sergeant for his cool handling of the situation and his complete report on Victorio's position. Based on that report, Colonel Grierson decided to march his main column "at a trot" through Bass Canyon to the Van Horn Mountains in an attempt to intercept Victorio.[104]

After midnight on the morning of August 5, the column received word that the Apaches had slipped by them crossing the stage road to the west of Van Horn Wells. Knowing that Victorio would head for the nearest permanent watering site, Grierson mounted his men and made a forced march sixty-five miles to the north. Company H lost four horses during the march. "Covered by a curtain of hills," a range later known as the Baylor Mountains, they arrived at Rattlesnake Springs in the Sierra Diablo range less than twenty-one hours later. Captain Carpenter had written of Rattlesnake Springs, "The water of this spring is always reliable, but is disagreeable to the taste, and not very beneficial for horses or men. In addition to the suphuretted hydrogen, so plainly tasted and to the smell, the water is also charged with some of the salts of lime, making it excessively hard." Nearby spread two large salt lakes. The ground between lakes and spring was flat, covered by sparse grasses and saline deposits. To the other side of the spring rose the jagged rocks of the Sierra Diablo. A wide cut wound back into the mountains from near the spring forming Rattlesnake Canyon.[105]

The Chase at Rattlesnake Springs. Print reproduced with permission of DonStivers.com.

At about 1:00 a.m. on August 6, after an exhausting ride, Carpenter's battalion rested and ate near Rattlesnake Springs. Among those leading Company H on the long passage north had been First Sergeant Toney Ratcliff, Sergeants George Garnett and Robert White and Trumpeter Silas Jones. About 3:30 a.m., Colonel Grierson arrived with the ambulances. Scouts reported the Indians camped at the south end of the Sierra Diablo near Fresno Springs. Grierson dispatched Captain Nicholas Nolan with Company A to scout nearby passes and water holes. After dawn, Captain Charles Viele commanding Company C and Carpenter's lieutenant, Charles Ayres, commanding Company G proceeded down Rattlesnake Canyon. Viele deployed his men at a high point in the canyon guarding the approaches to the spring. Finally at around 2:00 p.m., close to sixty Indians appeared coming down the canyon, still apparently unaware that they had been outflanked. As they advanced cautiously toward the spring, the cavalry opened fire. After some initial confusion, the Apaches realized that only two small companies opposed them and attacked in force.[106]

Before the soldiers could be forced from their positions, Captain Carpenter moved the rest of the battalion into place. Company H dispersed among the rocks, while Lieutenant Thaddeus Jones and Company B charged the Apaches driving them back into the surrounding hills and ravines. Around 4:00 p.m., Colonel Grierson's supply train under Captain John C. Gilmore and a detachment of Company H, Twenty-Fourth Infantry appeared on the scene from around the Baylor Mountains to the east. Seeing the loaded wagons, a dozen Apaches left the rocks to attack the train. The troopers soon forced them back under cover. The Indians managed to get a larger force past Carpenter's battalion and attacked the wagon train a second time. Soldiers of the Twenty-fourth, concealed in the wagons, leaped out firing and soon forced the warriors back into the hills. At that point Carpenter estimated almost 100 Indian warriors and ninety troopers facing each other on opposite sides of the canyon. He proceeded to form his men in a line of battle and advanced on the Apache positions. Reaching the rocks where the Apaches had been concealed, the soldiers discovered that most of Victorio's force already had retreated down the canyon. Horses were brought up as quickly as possible, but the Apaches had disappeared into the Sierra Diablo. The soldiers' pursuit was short lived. Many of the horses were still fatigued from the previous day's journey. Two companies returned to the spring within the hour and the other two after dark. During the battle, the Apaches had lost at least thirty dead or wounded. Company H had lost one man, Private Wesley Hardy from Georgia. One of Hardy's old comrades, Scott Lovelace of Company I later reported that Hardy had been carrying a message when the Apaches captured him, "I saw his saddle stripped of all leather. The Indians tied him to a tree and burned him alive."[107]

The following day, Captain Carpenter stationed his command further north at Sulphur Springs to prevent the Indians from continuing their move toward New Mexico, fortifying "the hills commanding the water hole, so that a small force could hold the position." On August 10, Colonel Grierson ordered Companies H and B to scout the Sierra Diablo for signs that Victorio was still in the area. The next day, Carpenter's troop found a trail sixteen hours old leading southwest. The captain immediately sent a dispatch asking Captain Nolan patrolling with Company A and

Company K, Eighth Cavalry to rendezvous at Eagle Springs. By this time, both horses and men of Carpenter's command were exhausted. They had been forced to abandon yet another four horses leaving them perilously short of remounts. Madison Bruin remembered having three horses during this period: "De fust one plays out, de next one shot down while on campaign and one was condemned." "While traveling over an unknown trail over sage brush, sand and alkali," John Casey later recounted, "we became very thirsty and very nearly perished from lack of water, as we had no fresh water, nothing but salty water for two days." Finally about 7 p.m. on August 11 "a cloud appeared on the horizon and in about twenty minutes the rain commenced pouring down in torrents." By midnight, Carpenter arrived at Eagle Springs. Rain had refreshed the weary troopers, yet they could not keep up the pursuit without rest. Nolan's command reached the spring about two hours later and continued on toward the Rio Grande. When Carpenter's men were able to cover the last thirty-five miles to the Rio Grande on August 13, Nolan reported that Victorio's band had crossed the night before with all of their women, children and animals. On August 20, Carpenter took Companies B, H and K and established a camp ten miles south of Ojo Caliente to patrol the border against Victorio's return. For a time at least, the Tenth Cavalry had succeeded in controlling the sources of water and driving the Apache raiding parties out of western Texas [108]

END OF THE INDIAN FRONTIER IN TEXAS 1880-1885

Captain Charles L. Cooper
Courtesy Fort Concho Library and Archives.

Following the fight at Rattlesnake Springs, Colonel Grierson wanted to follow the Apaches into Mexico, but the State Department had not yet received permission from the Mexican government or the state of Chihuahua. To ensure that Victorio did not reenter the United States, the Tenth Cavalry established an unbroken line of patrols all along the Rio Grande. From August to November, Company H covered a thirty-five mile stretch on either side of their camp near Ojo Caliente. Every other day their patrols would meet troopers from Captain Nolan's command covering the next stretch of river in the direction of El Paso. In September, Captain Carpenter received a report that prospectors had sighted several Indians along his sector of the river. On investigation, he discovered that the "Indians" were actually members of his own patrol from Ojo Caliente. The common practice of wearing slouch hats and comfortable clothing instead of the wool uniform while on patrol had confused the prospectors. The extended border patrols proved long and harsh for the men of Company H. For example, Romeo Satterthwaite of North Carolina, who had only been in service for one year became very ill with bronchitis and rheumatism from exposure. Carpenter sent him to Eagle Springs and then on to the hospital at Fort Davis. Satterthwaite remained in the hospital until November and continued to have problems with coughing and shortness of breath afterward. Sergeant Asa Weaver developed trouble with his vision while on duty. According to Sergeant

Robert White, Weaver's sight became so poor that he was relieved of guard duty "and had to be led about at night." When his enlistment ended the following year, his sight remained weak and he did not choose to continue what had seemed to be a very promising career.[109]

Throughout that period, Victorio's Apaches remained far south of the Rio Grande. United States troops eventually did receive permission from the Diaz government in Mexico to operate south of the border. In September, columns under Colonel George P. Buell and Colonel Eugene A. Carr marched into Mexico, to cooperate with Chihuahua state troops under Colonel Joaquim Terrazas while the Tenth Cavalry guarded the Rio Grande effectively blocking Victorio's escape route. The joint force located Victorio's camp in the Castillo Mountains in early October. Terrazas, who wished the honor of defeating Victorio himself, advised the American columns that their presence was no longer required in Mexico. Buell and Carr were forced to retreat to the New Mexico border. On October 15, Colonel Terrazas attacked and defeated the Apaches at Tres Castillos, killing sixty warriors including Victorio and eighteen women and children. Small handfuls of warriors and their families managed to evade the Mexican soldiers. In the aftermath of the battle at Tres Castillos, the detachment at Ojo Caliente commanded by Sergeant Charles Perry of Company B was surprised at daybreak on October 28 by about thirty Apaches. Those survivors of Victorio's band killed seven troopers including Corporal William Backers of Company H. They also captured enough horses and equipment to continue their flight into Texas.[110]

In early January 1881, other remnants of "Old Vic's" Apaches attacked a stagecoach in Quitman Canyon, killing the driver and one passenger. Texas Rangers from Company A of the Frontier Battalion stationed in Ysleta pursued the Indians north into the mountains of western Texas. The captain of the Rangers was George W. Baylor, a former Confederate colonel. He received command of the Ranger company after sending a letter to the governor of Texas in 1879 inquiring as to whether "he did not have some Indians he wanted scalped." Baylor had cooperated with the Ninth and Tenth Cavalry the previous year even following Victorio into Mexico on several occasions. Through snow and bitter cold, the ranger company continued to track the Apaches for over twenty days and 500 miles. They received reinforcements on January 24 from a detachment of rangers under Lieutenant Charles Nevill temporarily stationed at Fort Davis. The rangers pursued the renegades through the Eagle Mountains and into the Sierra Diablo. They finally overtook the Apaches near Alamo Spring deep in the Sierra Diablo range along the 1500-foot-high walls of the northeastern mountain slope. Early on the morning of January 29, they surprised the Indians, attacking in the pre-dawn light. Rangers killed four warriors, two women and two children. As the others scattered, Baylor recalled that his men sat down to breakfast amid "the ghostly form of Indians lying around." Like many others on both sides of the conflict, Baylor, normally described as a refined gentleman, could not consider Indians to be quite human. One Apache woman and two children were captured during the fight. Lieutenant Nevill took the prisoners back to Fort Davis along with an assortment of weapons and six cavalry saddles that the Apaches had taken from the soldiers in previous raids.[111]

Texas Rangers George W. Baylor and Charles L. Nevill

The skirmish with Baylor's Rangers proved to be the last Indian fight on Texas soil. While relative peace settled over western Texas, scattered raids continued in the New Mexico territory, primarily under the leadership of Nana, an eighty-year-old chieftain who had ridden with Victorio. On November 13, 1880, Company H had discontinued their patrols from Ojo Caliente and returned to Fort Davis. Colonel Grierson relinquished command of the now superfluous District of the Pecos on February 7, 1881. Looking back over the three years his command had been active, Grierson praised his troops asserting that "over one thousand miles of wagon roads and three hundred miles of telegraph lines have been constructed and kept in repair by the labor of the troops, a vast region thoroughly scouted over, minutely explored, its resources made known and wonderfully developed." Concerning Victorio, Grierson claimed that "his band of hostile Indians ... were outmarched, outmaneuvered, repeatedly headed off, disconcerted, met face to face, squarely fought, severely punished, driven into Mexico, badly crippled and demoralized, where - no longer able to hold together as an organized force - they fell an easy prey to the attack of Mexican troops and Indian scouts from the Sierra Madre." Departmental commander, General E. O. C. Ord also praised the Tenth's campaign against Victorio in his annual report to the Secretary of War "I trust that the services of the troops engaged will meet with that recognition which such earnest and zealous efforts in the line of duty deserve. They are entitled to more than commendation... In this connection I beg to invite attention to the long and severe service to the Tenth Cavalry, in the field and at remote frontier stations, in this department. Is it not time that it should have relief by a change to some more favored district of the country?"[112]

Relief did not come for the companies of the Tenth, still on patrol in the vast open lands of western Texas. In April 1881, the veteran troopers of Company H manned a sub-post near Presidio, Texas along with Companies F and G of the First Infantry. From here in May, they mounted a twenty-five-day scout into the Chisos Mountains of the Big Bend, returning after a month of fruitless searching for any further sign of the Apaches. During the summer, the troop stood guard while employees of the Texas and Pacific Railroad built a bridge over the mouth of the Pecos River. Much of the soldiers' duty in the field during this period involved guarding railroad workers and their supply trains. As the railroad companies began to recover from the Panic of 1873, plans were revived for a southern route of the transcontinental railway system crossing the open lands of Texas where only the year before the army had patrolled against the Apaches. The Texas and Pacific reached the future site of Abilene, Texas in January 1881. By March, lots were being auctioned for a new city to be named after the railroad town of Abilene in Kansas. A few months later a water stop was established at a site named Odessa for some of the railroad workers' home in Russia. At about the same time, workers set up camp at nearby Midway Station, halfway between Dallas and El Paso, that would become the city of Midland. By the spring of 1882, the rails reached the site of Alpine, only a little over twenty miles from Fort Davis.

While stationed out of Presidio, Sergeant Asa Weaver received his discharge as did Blacksmith Clinton Davis. James Jackson suffered a contusion to his right arm after being "kicked by a government mule, in the line of duty" and Dorsey Johnson sprained his left knee jumping off a horse. Albert Christopher, who had only arrived from the recruiting depot in February, could not join the company in the field due to an obstruction of the bowels. His condition continued to worsen and he died April 8, 1881 at Fort Davis. After leaving a small detachment at Presidio, the troop moved to Pena Colorado, near present Marathon, September 23, 1881. For most of the time they were stationed at Pena Colorado the company numbered only twenty-four to thirty-three men. Throughout this period, they continued scouting patrols deep into the Sierra Diablo and Guadalupe mountains. Company H returned to Fort Davis after another full year in the field on January 7, 1882.[113]

Captain Carpenter had been on one year's leave of absence in Europe since April 1881. He submitted a thorough report to the War Department concerning "equipments and troops abroad" and returned to duty with Troop H on April 25, 1882. Two weeks later, the troopers left on yet another patrol at Eagle Springs, watching the Rio Grande for signs of the few Apaches who still eluded the army. In December they returned to camp at Eagle Springs for one of the last times. While on that scout, William Allen, a former shoemaker who had become a minister during his tour of duty with Troop H, developed catarrh, a serious inflammation of the nose and throat, also suffering from increased deafness. He blamed exposure to the cold night air noting, "We had our tents and blankets, but there were no floors to the tents..." This proved to be a continuing and sometimes deadly condition. Returning from patrol in 1883, two Company H troopers, James Johnson and John Muchs suffered from an inflammation of the lungs and died of pneumonia at Fort Davis on August 2, 1883. Their bodies were shipped to departmental headquarters in San Antonio for burial.[114]

The year 1883 proved to be one of change for Company H. Cavalry companies whose members had long been referred to as troopers became officially designated as troops. In March, Captain Carpenter received promotion to major of the Fifth Cavalry, after sixteen years with H Troop. Charles L. Cooper, another Tenth Cavalry veteran, replaced him as commander of the troop dating from September 18, 1883. Cooper was born in New York on March 6, 1845 and stood 6'2". He joined Company B, Seventy-First New York regiment in May 1862 at the age of seventeen. He served in the defense of Washington, DC and was wounded at the battle of Gettysburg. On March 3, 1864 he received acceptance as a candidate for a commission with the United States Colored Troops and entered a training program at Taggart's Military School in Philadelphia. Cooper was commissioned a second lieutenant in the 127th USC Infantry on September 5, 1864. During the commissioning ceremony, a young woman named Flora Green held the presentation box containing a sword, sash and shoulder straps as they were given to the new lieutenant by his old New York unit. The two showed an interest in each other and according to a later account by one of their children, Flora sent Charles Cooper back to the battle lines with flowers the next day.[115]

Cooper's first command soon found themselves on the front lines during the final encounters of the war. On March 27, 1865, they captured rebel fortifications at the second battle of Hatcher's Run. In front of the Confederate defenses at Petersburg, on the afternoon of April 2, Cooper led his company "in the often spoken of attacks on the rebel fortifications known as Fort Damnation" advancing "in the thickest of the conflict under a most destructive fire from the enemy." The next morning, they entered Petersburg and Cooper received a brevet promotion to captain. Six days of forced marches followed before they finally overtook the Confederates on April 9 near Appomattox. There the southern forces attempting to break through Sheridan's cavalry were "brought abruptly face to face with the colored troops just arrived...they faltered and soon hastily retreated."[116]

Following the Confederate surrender, Cooper's regiment transferred to the Mexican border as part of a show of force against the French government of Emperor Maximilian. While there Cooper met a board for commission and appointment to the regular army. A rupture in his right groin caused him to fail the physical. He returned to Philadelphia in November and married Flora Green on December 20, 1865. Cooper briefly tried his hand running a newspaper, but civilian business did not suit him. In July 1866, he received an appointment as second lieutenant in another Black regiment, the Thirty-Ninth Infantry. During 1870, while serving as Indian agent in Santa Fe, New Mexico, he requested a transfer to the cavalry. Cooper received his new assignment on the last day of 1870, joining Company A of the Tenth Cavalry at Camp Supply, Indian Territory. With Company A, he served at Fort Sill during the Red River War and later at Fort Concho. Flora and Charles had three children during this period, Forrestine, called "Birdie," Harry and Florence.[117]

In July 1877, Lieutenant Cooper was a member of the tragic Nolan expedition across the Staked Plains of Texas. Captain Nicholas Nolan had set out from Double Lakes on the hot, dry afternoon of July 26, following the trail of about forty Indians. Although he was a veteran cavalryman, Nolan's wife had died in February leaving two young children and it would appear that he still suffered greatly from the loss. He apparently did nothing to ensure that his company filled their canteens or carried extra water. By the next evening, following thirty hours without fresh water, many of the command suffered from heat exhaustion. Cooper later wrote his father that of "forty rational men who left camp with us" after two days "our party now consists of eighteen madmen." He recalled that by July 28, the men's "tongues were swollen, and they were unable even to swallow their saliva - in fact they had no saliva to swallow" while the exhausted troopers were "fighting each his neighbor for the blood of the horses as the animals' throats were cut. . . My tongue and throat were so dry that when I put a few morsels of brown sugar that I found in my coat pocket into my mouth, I was unable to dissolve it in order to swallow it." On July 29, Cooper urged a new course on Captain Nolan telling the men to travel by night carrying only their weapons and "endeavor to reach some of the streams to the east of us." Finally at around 5:00 a.m.

on July 30, the first members of the straggling column reached Double Lakes. They had been without water for eighty-six hours.[118]

Although Cooper's calm actions helped bring his men safely back to camp, they did not gain him any favor with Captain Nolan. Cooper and Nolan seem to have disagreed frequently. In 1879, Nolan formally reported him "for failure to forward personal reports." Cooper managed to transfer out of Company A in 1882 when he became regimental adjutant for the Tenth Cavalry as the headquarters moved to Fort Davis. The following year, he requested appointment as a captain in the quartermaster department. Captain Carpenter wrote him a letter of recommendation. Instead of being assigned to the quartermaster department, Cooper's next assignment was to replace Carpenter in command of Troop H in September.[119]

Colonel Grierson had moved the headquarters of the Tenth Cavalry to Fort Davis in July of 1882. As settlement advanced on the Texas frontier, for one of the first times since the company's formation, Troop H spent an extended period performing routine duties on post. Members of the troop received various special assignments that included teamster, headquarters clerk, post gardener, hospital cook and "Stable Police." Others performed extra duty for the officers' families like Michael Finnegan who sometimes acted as Birdie Cooper's bodyguard. By 1884 one man out of every six had served more than one enlistment. Desertion and alcohol addiction in the Tenth Cavalry were among the lowest in the service. Soldiers did still desert, but it was usually new recruits unaccustomed to the isolated conditions on the Texas frontier like the blacksmith William Nugent who enlisted in May 1884 and deserted in August. Some of the old soldiers like Sergeant Pollard Cole even married and established a home at Fort Davis. Wives of married soldiers frequently worked on post as laundresses or servants for the officers' families. Jenny Miller, the wife of Sergeant Girard Miller worked in the Coopers' home.[120]

Private Andrew Emery had an opportunity to work with Chaplain Francis Weaver to improve the post library which by that time had exceeded 1000 volumes. Weaver served as chaplain of the Tenth Cavalry and had moved to Fort Davis with the headquarters in 1882. Under army regulations, only posts in the most isolated areas were allotted chaplains; however, each of the four Buffalo soldier regiments was to have their own chaplain. Chaplain Weaver, unlike many of those assigned, had fought as an enlisted man during the Civil War being wounded twice at Fredericksburg and three times at Gettysburg. A native of Pennsylvania, he graduated from Gettysburg Lutheran Theological Seminary in 1876 and served as a missionary to the Southern Utes in 1877 and 1878 before joining the chaplain service in 1880. In a paper he wrote concerning frontier service he stated the belief that "more persons proportionately attend service in the military than in civilian life." Weaver enjoyed music and believed that it was important to include singing in church services and Sunday School. In addition to religious services, he worked as post treasurer and librarian. He also continued the work of his predecessor at Fort Davis, Chaplain George G. Mullins, by improving the quality of schooling for both the children and the enlisted men.[121]

The Army Reorganization Act of 1866 provided "That whenever troops are serving at any post, garrison, or permanent camp, there shall be established a school where all enlisted men may be provided with instruction in the common English branches of education, and especially in the history of the United States ..." The law also called for the post commander to ensure that a building was allotted for conducting school. By the 1880s Fort Davis had a chapel that could seat over 200 people and could also be used for school. Although many of the troopers did not attend school regularly after the long hours of fatigue duties and drill, those who did benefited directly. Education made them more valuable as non-commissioned officers capable of handling the complex army paperwork and it gave them a better opportunity for civilian jobs after their enlistment. While the study of United States history was unquestionably important for those who were serving and defending the country, it may have had an adverse consequence. One history textbook of the time described Native Americans as "distinguished by a remarkable want of foresight" and watching "every movement around them with suspicion." It went on to say "Strongly attached to their savage mode of life, they will not give it up until obliged to do so." Another text speaking of troubles with the Creek nation during the early nineteenth century recounted that the Creeks and their allies began "to commit outrages, in the usual Indian fashion, upon the families on or near their borders." Combined with their personal experiences tracking down Indian raiders, such instruction probably reinforced the Black troopers' view of Native Americans as irrational savages rather than another minority group deserving their respect and understanding.[122]

Life at Fort Davis also included occasions where the only purpose was to enjoy the company of fellow soldiers. The *Army-Navy Journal* described New Year's Day of 1884 on the post as "brilliant with sunlight like a clear October day in the north." Every troop of the Tenth Cavalry hosted a dinner, decorating the mess hall tables in red and white. Preparations for the feast lasted all day involving the troopers in activity far more pleasant than the usual drill or patrol. "Vast fires were burning in the Troop ranges, and odors of cooking game, turkeys, chickens, pigs, cakes, pies, puddings, sauces and all else known to be good to sustain the weary frame of the soldier..." A formal ball began at 9:00 p.m. After the initial round of dancing, the great feast was served at midnight. More dancing followed until 2:30 a.m. when the band finally ended the celebration with the strains of "Home Sweet Home." On Christmas eve of that same season, Captain Cooper, Captain Keyes and their families had attended midnight mass at a small adobe church across the border in Chihuahua.[123]

Such feasting remained a rare occurrence and the normal army ration with few fruits and vegetables continued to take its toll on the soldiers' health. Sergeant Thomas Maddox, a native of Texas with nine years' service, died on June 1, 1884, from an abscess of the liver while on duty at Fort Stockton. Benjamin Banks with less than two years in the cavalry at Fort Davis died there of acute dysentery on November 25, 1884. For those in good health without the rigors of riding long patrols, the quiet life on post may have seemed uninteresting after so many months in the field. In October 1884, First Sergeant John Casey decided to go hunting with Lance Sergeant Robert Anderson of the regimental band. Obtaining a pass and a government wagon, they left the post after inviting two prostitutes, Pearl Wilson and Rachel Hall, to join them on the hunt. The troopers stayed in the mountains with the two women for three days before returning to post. When stories began to circulate around Fort Davis, Sergeant Casey, whose wife Pabla and two children were living with him in government quarters, became rather nervous and evasive. Captain Cooper questioned Casey about the expedition, but he denied that the prostitutes had been with him. He even told Chaplain Weaver that "enemies were endeavoring to injure his character." Because of this unusual behavior, Casey was brought before a court martial for deceiving his company commander and the chaplain. During his trial, the judge stated, "If there is anything that the First Sergeant ought to do it is to tell the truth to his Company Commander." John Casey was found guilty, reduced to private and ordered to forfeit ten dollars a month for six months.[124]

Casey did not need to worry about breaking the monotony of post routine for much longer. After ten years in the Trans-Pecos region, Troop H rode out of Fort Davis for the final time on April 1, 1885. With the rest of the Tenth Cavalry, they had been assigned to outposts in the Arizona territory. There some of the Chiricahua Apaches under Geronimo and Mangus as well as the remnants of the Warm Springs band under Nana still left the reservations to pursue their traditional way of life. During their ten years at Fort Davis, Company H had compiled an enviable record of service in the field. Old soldiers like Silas Jones and Joseph Claggett proudly tattooed the letter "H" on their arms. In 1877, Chaplain Mullins of the 25th Infantry had written from Fort Davis, "The ambition to be all that soldiers should be is not confined to a few of these sons of an unfortunate race. They are possessed of the notion that the colored people of the whole country are more or less affected by their conduct in the Army." That observation could certainly be applied to Company H during this period. Since their initial organization in 1867, the company had taken part in every Indian campaign on the southern plains serving at outposts from Kansas to Texas. Patrolling thousands of miles over numerous unknown trails, the men of H Troop had become a crack outfit and a compelling example of the dedicated service given by Black soldiers on the American frontier.[125]

BUFFALO SOLDIER BLUES: DAILY LIFE IN H TROOP

The Promise. Print reproduced with permission of DonStivers.com.

As Troop H moved from Kansas to Texas to Arizona, despite changes in climate and terrain, daily life on the outposts of the U. S. Army remained much the same. Their day often began as early as 5:00 a.m. when the soldiers awoke to the sounds of a "morning gun" and the bugles blowing reveille. Before the cavalry troopers ate breakfast at 6:00 a.m., they went to the stables to feed and groom their horses. At 6:30 a.m., any soldier unfit for duty reported to the hospital for sick call. After breakfast, the garrison assembled to post the guard. Officers inspected each company and their quarters. At 8:00 a.m., those soldiers not on guard assembled to perform various fatigue duties including the cleaning and repair of equipment and buildings, hauling wood and supplies, and care of the animals. On army posts that were still being built, duties might also include quarrying stone and the actual construction of buildings. In Troop H a common rotation of duties included post guard, stable guard, general fatigue, kitchen police and room orderly. The largest meal of the day was served around noon. Soldiers then returned to fatigue and guard duties for several hours followed by military drills and target practice. At dusk, retreat sounded and supper was served. In the summer, the day ended around 8:30 p.m. During the winter, with limited daylight, the soldiers' day ended even earlier.[126]

At all of the posts along the frontier, a similar routine marked the pace of a soldier's life. Off duty soldiers could work as strikers which involved acting as a man servant for one of the officers. This was considered a desirable job, because they worked in comfortable quarters and frequently received extra food and money. The other soldiers frequently called them "dog-robbers" because they ate the extra food that might have gone to the family dog. The average trooper received a daily ration that consisted mainly of beef, bread and coffee. Even at breakfast, coffee, beef and bread remained the standard fare. For supper, soldiers ate food warmed over from the large noon meal. In the common folklore of the Black regiments, the words were put to mess call that included, "Soupy, soupy, soupy, widout a single bean, Po'ky, po'ky, po'ky widout a streak ob lean." Whenever possible, cooks supplemented the unappetizing ration with potatoes, bacon, hominy, bean soup and even fresh fruits and vegetables. Troopers such as Randall Blunt, on detached service, cultivated gardens on post with some success. Officers would pool company funds to provide occasional delicacies. Some of the foods bought to supplement the unappealing ration included clams, sardines, salmon, dried apples, dried peaches, raisins, prunes, pickles, jellies, preserves, canned milk, tomatoes, green corn, green peas, potatoes, asparagus, onions, peaches, pineapples, cranberry sauce, oysters, and various spices. Soldiers returning from hunting trips sometimes brought fresh game including buffalo, antelope and turkey. Yet the supply of fresh vegetables and items that made the meals palatable such as butter, honey and lemons remained scarce at all of the frontier outposts.[127]

On parade and on guard duty the soldiers wore wool cavalry jackets and trousers, with high-topped cavalry boots. A writer to the *Army-Navy Journal* complained, "The heavy woolen clothing which the rank and file of our army are condemned to wear under the burning rays of a Midsummer sun, whether upon the Western plains or in the Southern portion of our country... I would simply ask as a special favor to poor, sweltering, melting, perspiring humanity...that we might obtain the boon of 'lighter clothing' during Summer." Whenever possible the troopers would discard the heavy uniforms. Stable duty allowed them to wear canvas stable frocks and overalls. When on patrol, the troopers frequently wore non-uniform items, especially broad brimmed hats which were not available through army issue. Birdie Cooper recalled officers returning from a scouting patrol in blue flannel shirts, "Mexican sombreros, unshaven faces, and Indian moccasins." During the frozen winters, heavy clothing became an advantage, especially the army greatcoat with cape. In January 1885, Scott Cain, who had sold his greatcoat in violation of regulations, entered the barracks during a formation to look for his "Mexican hat" and walked out with Peter Dehoney's coat. Cain, described by Dehoney as "a peculiar little man," was court-martialed for the theft and for being absent from formation without permission. He received a dishonorable discharge and confinement at hard labor for one year.[128]

Most of the soldiers in Troop H ranged in age from twenty to thirty-five. Twenty-one proved the most common age for enlistment. Many registered their occupation prior to enlistment simply as laborer. Others had worked as farmers, shoemakers, barbers, blacksmiths and waiters. Over sixty percent of Troop H listed their place of birth as states along the border of the former Confederacy particularly Kentucky, Maryland, and Virginia. Most from the South claimed Mississippi or North Carolina as their home. Many were uncertain of their date of birth, having been born in slavery. Few of the troopers came from northern states other than Pennsylvania, although some did come from the Midwest and several were born in Canada. This proved to be very different from other regiments such as the Second or Seventh Cavalry where most of the troopers came from the North and Midwest and forty to fifty percent were foreign born. Very few members of Troop H had ever lived in any of the states or territories to which their military service took them. In fact, they were frequently the only Black inhabitants in the county or region. Sparsely settled Presidio County, Texas claimed 1009 male citizens in 1880. Of those, 429 were Black, almost all of them stationed at Fort Davis.[129]

In 1877, Chaplain George M. Mullins of the Twenty-fifth Infantry wrote from Fort Davis where Troop H was stationed, "The ambition to be all that soldiers should be is not confined to a few of these sons of an unfortunate race. They are possessed of the notion that the colored people of the whole country are more or less affected by their conduct in the Army." Mullins, a Disciples of Christ minister who had initially been reluctant to work with Black troops continued, "Their interest in school is unabated, and upon public Divine service the attendance large." Secretary of War Redfield Proctor wrote of Black troops in his annual report: "They are neat, orderly and obedient, are seldom brought before court-martial and rarely desert." Captain John Bigelow pointed out that during the period the army experienced 2811 desertions during one fiscal year for an average of seventy per unit. At the same time, the Tenth Cavalry had only twenty-three deserters. Birdie Cooper praised the troopers, claiming, "All the enlisted men of the Tenth Cavalry were colored soldiers of the best type. [They] endured untold hardships, fording dangerous rivers, traveling where the only water obtainable was unfit for use, facing Texas northers, withstanding hailstorms one hour and sweltering heat the next. 'Just in the line of duty,' they said, 'and all in the day's work.'"[130]

Captain Louis Carpenter provided an interview to the Omaha *Bee* in April 1883 concerning the character of those troops he had until recently commanded. Asked what kind of soldiers Black men made, he replied, "A great deal better than many suppose... The records show that there have been less desertions in the Tenth cavalry than in any other regiment in the service. We have had fewer court martials, fewer offenses against the regulations, and as good general discipline as can be found anywhere in the army. The men are unusually cleanly and tidy, and spend more on their dress than the white soldiers; they drill well, and are obedient to their superiors." Commenting on their conduct under fire he stated, "My experience has been ... that they are as reliable as white soldiers in action. I have seen them in a number of Indian fights, and they behaved unusually well... In '68 the two companies that I commanded did excellent service and deserved all the compliments that they got." Yet even after almost twenty years in command of Black troops, Carpenter still had some misgivings concerning their enlistment. "The moral tone of the colored men is not as high as that of the whites," he asserted. "They are more uneducated and it is extremely difficult to find mechanics among them for the work of the regiments." Although Carpenter thought enlistment of Black soldiers should have waited until they were better trained, his own experience led him to conclude, "Still it is only fair to say that they are improving greatly in the matter of education and will doubtless improve more rapidly in this respect every year.[131]

Even while they learned the skills necessary for advancement, the troopers continued to endure the hardships of posts that were always on the frontier of civilization. Only a few of the senior enlisted men had families to greet them when they returned to post. Army regulations prevented married men from enlisting, yet soldiers could marry on active duty with the permission of their commanding officer. Many of the enlisted men's wives worked as servants for the married officers' families or as company laundresses. Employing non-commissioned officers' wives as laundresses became so common on the Texas frontier, that in 1876 General Edward O. C. Ord declared that many of his best soldiers would be unable to re-enlist if their wives were not also provided work. Due to the severe conditions on many posts, injuries occurred much more frequently on routine duties than when in pursuit of hostile Indians. George W. Forster, who had been wounded in action in 1879, had a crowbar fall on his hand while laboring on post and then sprained his right knee when a horse fell with him in February 1883. By September 1883, he had developed acute rheumatism, requiring regular treatment. On February 20, 1891, Forster was shot in a dispute with interpreter John Glass at Fort Apache. He died two days later after twenty-one years of faithful service and a variety of injuries. John Morgan also died of a gunshot wound at Fort Davis; however, it proved to be an accident. He had been in the hospital with a cough until that morning and then returned in the evening, shot in the abdomen when a weapon discharged in the barracks.[132]

Sprains and contusions seemed to be the most common results of accidents on post and in the cavalry, many of those accidents were caused by horses. Charles Barnum, Dorsey Johnson and James Andrews all injured themselves falling from their horses. Of course, James Andrews suffered a more serious concussion in December 1880 when another soldier beat him on the head with a club. Washington Hardaway, while on herd duty near Eagle Springs, felt his horse stumble and then raise itself up violently.

Hardaway hurled forward onto the pommel causing a hernia from which he did not recover, receiving a disability discharge on September 21, 1880. Corporal William Webb, wounded in action in 1879, was tying down a horse when it reared up and kicked him in the head just above the right eye. Blacksmith Henry Walker had a horse fall on his foot and a few months later he smashed his finger with an axe. In November 1884, while working in the blacksmith shop at the company corral, a piece of iron struck Walker in the right eye. He lost all use of that eye and received a disability discharge in March 1885.[133]

If horses could be dangerous even to cavalrymen, pack mules were worse. James Jackson suffered a contusion of the right arm at Pena Colorado in 1881 being "kicked by a government mule, in the line of duty." Another mule had kicked Private Colonel Miller in the left knee a few months before. At the time Jackson was kicked, Miller was being treated for a back injury received while playing baseball. William Webb strained the cords in his right hand on saber drill. His wrist became badly swollen and his fingers would only bend half way. William Hawkins, a tailor by trade, suffered sunstroke during target practice at Fort Davis in the summer of 1883. He later wrote, "We had been laying out there in the sun every day for two or three weeks." At Fort Sill in 1870, Bartlett Mates accidentally discharged a shotgun that he believed to be unloaded. The blast hit Charles Black in the right eye. Black's vision in that eye was destroyed and Captain Carpenter noted "the soldier has been entirely unfit for duty ever since." Black received a disability discharge on March 28, 1871.[134]

Diarrhea and dysentery were among the most common diseases, largely due to the limited diet of beef, beans and coffee. Almost twice as many soldiers reported to sick call for diarrhea as for any other single disease. Benjamin Banks died of acute dysentery on November 25, 1884 at Fort Davis. Scurvy also became a problem at times because of the lack of fresh fruits and vegetables. Charles Black and Bugler James Thomas suffered from scurvy during the winter campaign of 1868. Problems due to poor diet seem to have affected every frontier outpost from Dakota to Texas with post surgeons lamenting the shortage of fruits and vegetables available to the soldiers. Surgeon Samuel Smith noted that he felt like he was "boarding at a first class dyspeptic factory" referring to the numerous cases of indigestion at Fort Concho while Doctor Washington Matthews at Fort Rice on the Missouri River in Dakota Territory remembered, "During the first years, scurvy was a formidable malady, destroying many lives, and otherwise seriously reducing the efficiency of the garrison."[135]

After many miles in the saddle and many nights on the hard ground, rheumatism also proved to be one of the most common ailments. Troopers attending sick call for rheumatism became even more prevalent than those suffering from coughs and fever. Other familiar ailments included bronchitis, cattarh, tonsillitis, headaches, fever, mumps, boils and piles. Sexually transmitted diseases seemed to always be present among the mostly unmarried soldiers. Even thirty-year veterans like Joseph Claggett had to battle syphilis. Trumpeter Silas Jones, present in almost every Troop H campaign had gonorrhea in 1877 and syphilis in 1880. There were also soldiers who appear to have suffered constantly from one ailment or another. James A. Hill enlisted at Davenport, Iowa in August 1888. On April 26 of the following year, he collapsed in inspection ranks and had to be carried to the hospital. On May 23, he again suffered vertigo while standing on the porch of his barracks. Hill continued to spend time in hospital with headaches and eventually was deemed "unfit to perform the duties of a cavalry soldier" and discharged on January 27, 1890 at Fort Apache. Over a two-year period, William C. Alexander was treated for bronchitis, colic, fever, catarrh and rheumatism on a recurring basis. He finally received a certificate of disability and was discharged on August 22, 1874.[136]

Sometimes disease killed the Black troopers even more surely than Indian weapons. Two Company H troopers died of pneumonia at Fort Davis in 1883, James Johnson and John Muchs. Muchs had served in H Troop for almost ten years sending part of every paycheck home to his mother in Kentucky. Although illiterate, he had others troopers write notes that he sometimes included with the money. George Johnson died of pernicious malarial fever at Fort Sill in July 1872, while John Lisby perished of typhoid on June 29, 1877. An inflammation of the lung killed Dorsey Johnson on March 5, 1883. Johnson had enlisted at Louisville, Kentucky in 1881 at the age of nineteen. He previously had been employed by a tobacco factory in Louisville and according to his mother, "He enlisted with my consent: he said he thought it was the best thing he could do." Dorsey's mother, Georgia, barely managed to survive working as a chambermaid on Mississippi steamboats. He would send her two, three or five dollars from each army paycheck. After his death, Georgia claimed and received a pension as a military dependent.[137]

Tenth Cavalry at Fort Davis. Courtesy Fort Davis National Historic Site.

For soldiers who avoided accidents and illness, the regimented life of the frontier army, where strict adherence to orders could mean the difference between life and death, sometimes became too great a challenge and they faced military justice in front of a court-martial under the Articles of War. Although Troop H maintained an exemplary record throughout their period of active service, some of the case records from the Office of the Judge Advocate General in the National Archives demonstrate the types of offenses, from fatal to almost comical, that affected the troopers' daily routine. As always on the frontier, desertion continued to be one of the more common court-martial offenses. Sometimes there were even unwilling deserters. Private Alexander Brown had stolen weapons from the ordnance room to aid in his escape. Deciding that he did not wish to leave alone, he offered Robert Miller $16 and three revolvers to desert with him. Miller, testifying against Brown at his court martial declared, "He asked me was I going to desert with him. I told him, yes. He had a pistol in his hand cocked."

Brown received a dishonorable discharge and confinement for five years. Miller returned to duty.[138]

In other cases, a record of smaller offenses eventually led to desertion. Private Anderson Wilson "while a member of the stable-guard of said troop, was duly posted as a sentinel, and did sleep upon his post." at Camp Supply in 1870. He received confinement at hard labor for six months and was fined $10 per month for the same period. On another occasion when Lieutenant Thomas J. Spencer told him to return to his guard post "He sauntered off slowly with his hands in his pocket..." Wilson continued to be a discipline problem. Finally on June 21, 1872, while on patrol, he managed to shoot himself in the hand when picking up his gun. Reuben Waller, now employed as a civilian teamster, observed the sergeant give Wilson permission to go to the rear of the column. Wilson decided not to stop when he reached the end of the column. He later testified, "Knowing that there was no surgeon along, and knowing that I would have to go about five hundred miles with the command before I could see one, I thought I would go back to Arkansas City." Troopers finally caught up with him about seven miles southwest of Arkansas City on road from Fort Sill. Wilson received a dishonorable discharge and confinement for two years in prison at the Texas State Penitentiary in Huntsville. Conviction for desertion almost always received a punishment that included confinement in prison and a dishonorable discharge. The maximum prison sentence for such an offense was five years, yet few soldiers received the maximum. The most common length of confinement was three years.[139]

Another unusual case of desertion was Frank Posey who enlisted in Company H on November 21, 1881 at Baltimore, Maryland. He deserted in February 1885, just a few months before the troop left Fort Davis. He later testified that local officials had threatened to arrest him and "thinking that I had served so faithfully up to that present time, it would be better to desert than to lose my military record by being in the hands of the Civil Authorities," Posey deserted. Still preferring the army life, Posey proceeded to enlist in the Twenty-fifth Infantry under the name, Frank Brocko. In May 1888, while serving with distinction at Fort Snelling, Minnesota, he was recognized as a deserter, arrested and brought before a court martial. Posey's sentence included a dishonorable discharge, forfeiture of all pay and confinement for two years at hard labor. Based on his service record, under both names, the court granted clemency and reduced the sentence to one year of hard labor at Fort Snelling. They later commuted the sentence following a request from the governor of Minnesota stating that Posey had a wife in Saint Paul "who is sickly and is in destitute circumstances."[140]

Other soldiers resented the harsh army life and sometimes retaliated against those placed in authority over them. When Quartermaster Sergeant Scurry told John H. Curtiss to help feed the horses, he refused, insisting "I was not drove by a black man when I was out of the service and I will not be while I am in it. I will not be drove by any of your kind." Curtiss was sentenced to two months hard labor and ordered to forfeit $8 per month for the same period. In August 1888, while on duty at the San Carlos reservation, William Epps received orders to assist in digging a sink for the detachment. Instead, he wandered away and did not return until other members of the company had finished the work. As he sauntered back toward the workers, Lance Corporal Randolph Nelson ordered him to return a shovel borrowed from the Twenty-fourth Infantry. Epps refused and was sentenced to hard labor for four months and forfeiture of $10 per month for the same period. Doc Mocfield was drunk and disorderly at Fort Apache in 1887. Corporal George Forster ordered him to halt, while Corporal Michael Finnegan attempted to arrest him. Mocfield proceeded to beat on Finnegan and curse him until the corporal finally got him to the guardhouse. Mocfield was confined at hard labor for four months and forfeited $10 per month during that period.[141]

Drunkenness continued to be a problem at many of the frontier outposts. On the afternoon of December 8, 1875, Private Greenfer Shanklin failed to appear at water call and stables. He had been drinking heavily. When Sergeant James Campbell came to escort him to the guardhouse, Shanklin grabbed his carbine and threatened to shoot Campbell or anyone who tried to take him. Lashing out wildly, he finally broke the carbine in an attempt to hit First Sergeant George Garnett who had responded to the disturbance. Shanklin received a sentence of dishonorable discharge and confinement in the penitentiary for two years. Benjamin Harris, while pleading guilty to a charge of public drunkenness, stated that though he could remember nothing, "I feel deeply the disgrace to which the excesses of drink has brought me...the result of unfortunate weakness to which the flesh is heir..."[142]

While Harris showed remorse for his actions, others like Sergeant Charles Gray showed only anger. On October 1, 1889, Gray and First Sergeant Michael Finnegan walked over to a dance in the Troop A dining hall at Fort Apache. Looking into the hall, Gray noticed Frederick Plotten, a civilian employee of the quartermaster department. He believed Plotten had been spreading rumors that Gray had stolen money, because he was carrying a large roll of cash on the day of the last baseball game. So Gray sent a message to Plotten asking him to step outside and settle the matter. Plotten believed that First Sergeant Finnegan had called him out and was shocked when Gray confronted him with the stolen money accusation. He denied spreading such a rumor, but Gray only became more angry and began to hit Plotten. Finnegan tried to separate them. A crowd quickly surrounded them, and in the confusion Plotten escaped. Gray chased after him, while Jake Smith of the regimental band, thinking that Plotten was escaping from First Sergeant Finnegan, caught and held the civilian. Finnegan finally managed to calm Gray and explain the confusion to Smith and all of the soldiers returned to the dance. Yet as he walked back to the dining hall, Finnegan later asserted, "I could hear some say that Sergts Gray and Finnegan had beaten the packer and some saying that Sergt Finnegan had been trying to stop the fight. It was such a mixed up affair..." Plotten, still confused by the circumstances, charged both Gray and Finnegan with "violent and murderous assault." In the resulting court martial, the matter was finally sorted out and Finnegan was acquitted of any part in the fight. Gray was reduced to private and ordered to forfeit $10. It proved extremely fortunate and probably a tribute to his effectiveness as a soldier that Gray did not spend some time in confinement or at hard labor. He had previously been convicted of verbally abusing Private Lewis Thomas for using his comb. In addition, the courts generally did not like to forfeit pay if the soldier was not also confined, because they believed a soldier doing full duty should receive full pay. The reduction to private, a common sentence for non-commissioned officers, was probably thought to be more effective than confinement.[143]

Like Gray, other non-commissioned officers sometimes failed to follow the regulations they had been charged to enforce. At Camp Supply in 1870, Sergeant James Clayton as sergeant of the guard allowed Private James Henry of Company A to escape from the guardhouse. For this mistake, he forfeited $5 per month for three months. At Fort Wallace in December 1868, First Sergeant Amos Cormack gave permission for Robert Edwards and John Brown to go into Pond City after orders had been issued that no passes should be granted for the town. While brought to trial, he easily was acquitted, having had no knowledge of the orders. Then there was Sergeant Henry Allen. On October 4, 1873, Allen was on duty as sergeant of the guard. He returned to the guard post half an hour after he should have reported back from supper, allegedly because the meal was not ready on time. When the officer of the guard, Lieutenant Robert G. Smither, asked where he had been, Allen, with his pipe stuck in the corner of his mouth and leaning against a banister, insolently replied, "Well I guess I was eating my supper." He refused to stand at attention or go to his quarters. When Smither demanded that he surrender his weapon, Allen slid his gun across the ground instead of handing it over. Smither then drew his sabre and forced Allen into the guardhouse, where the sergeant grabbed the officer around the waist and attempted to drag him into the cells. At Allen's court martial, Captain Carpenter testified to his excellent character as a non-commissioned officer, but the prisoner was still convicted, dishonorably discharged and confined to a military prison for one year.[144]

Yet Allen's sentence was overturned and he returned to duty, because he had not been allowed to testify in his own defense in violation of the rules for General Courts-Martial. In overturning the verdict, the court admonished, "The conduct of this prisoner was insolent and insubordinate in the extreme and it is to be hoped that in shaping his future conduct he will remember that he did not escape punishment on account of any justification he had, but by reason of a flaw in the proceedings, an accident which may not occur again." But Allen did not learn. In February 1876, he was in trouble again for being absent from retreat "in consequence of his being drunk in barracks." Released from the guardhouse for target practice on February 7, Allen went to see Captain Carpenter demanding to know why he was in arrest. After the previous court martial, Carpenter would grant no leniency. He insisted Allen "should behave as a soldier in every respect or else I'd have him punished." Allen was ordered to return to his quarters, but instead he left the post. Carpenter sent a detail under First Sergeant George Garnett to capture him. Garnett found Allen concealed behind a rock, about a mile and a half from the post. Allen threw a rock and cussed at the first sergeant, insisting that he had taken nothing with him and did not intend to desert. He later insisted that Garnett would have killed him if there had not been witnesses. This time Allen was convicted of being absent without leave, sentenced to forfeit $10 per month for six months and confined under guard for the same period. Despite continuing attitude problems, Henry Allen served for thirty years in Troop H, retiring in 1897.[145]

Another thirty-year veteran who committed a court martial offense was Silas Jones. On April 23, 1870 at Camp Supply, he fell asleep on duty while guarding the hay stacks. Captain Carpenter defended Jones, testifying, "He has always been a good soldier and performed his duty well." Jones was released without punishment. Pollard Cole, another veteran who did not fare as well, was found sleeping on his post at Fort Davis in 1879. Under oath, Cole admitted, "I heard him cry half past one; it was the last hour I heard him cry." Sentenced to be confined in the guardhouse for four months and forfeit $10 per month for that period, Cole's sentence was later reduced to two months. William Hamlin received a dishonorable discharge for sleeping while on stable guard at Camp Supply. Private George Bumpferts found asleep on stable guard by Captain Nicholas Nolan during the same period was ordered to forfeit $10 per month for six months and be confined under guard for same period with the statement, "The court is thus lenient on account of the previous good character of the prisoner."[146]

Private Robert Ellis, also found sleeping on duty at Fort Apache in September 1891, received a sentence of four months hard labor and forfeiture of $10 per month for the same period. On February 1, 1892, while still serving a sentence of hard labor in the guardhouse, Ellis received orders from the sergeant of the guard to light the street lamps on post. He ignored the order, later explaining, "It was a moonlit night and the prisoner did not suppose the Sergeant intended what he said when he directed the lights to be lit." On trial once again, Ellis claimed he had already served seven months in the guardhouse when he had only been sentenced to four months. The court had no sympathy with his insubordination, sentencing Ellis to an additional two months of hard labor and forfeiture of $10 per month for that period.[147]

Another case of insubordination occurred as a work detail returned from watering the horses at Camp Supply in 1870. Ephraim Smith lingered behind at the water's edge. He later claimed to be rounding up two of the horses that remained loose. First Sergeant Jacob Young ordered Smith to close ranks. Smith ignored him and went directly to Captain Carpenter with his story. Carpenter listened to both sides of the story and then ordered Smith to guardhouse. Sergeant Charles Burns noted that Smith cursed the First Sergeant all the way to guardhouse saying "you are a damned black son of a bitch, and he will fix you." For his insubordination, Smith was confined at hard labor for one month. In a less serious occurrence, Private Dickson Hunter was caught talking to prisoner Robert Miller at the company sink at Fort Davis in June 1876. When sentinel William Jackson tried to stop him, Hunter cursed at Jackson and tried to disarm him. In that case, the court did not find sufficient evidence to convict him. In yet another case, Charles Jefferson lost his new break open Smith & Wesson revolver. While saddling horses for a scout from Fort Quitman on October 25, 1875, one of Jefferson's buddies playfully pulled out his pistol and did not get it back in the holster well. Jefferson proved to be guilty of nothing, but carelessly failing to check his holster before riding out and he was swiftly acquitted.[148]

While Jefferson's loss was innocent, that did not prove to be the case with John Henry. Having lost his carbine at Camp Supply in 1870, Henry stole William Bright's weapon. When it appeared he might be caught, he defaced the carbine to keep it from being identified. That ploy failed and Henry was dishonorably discharged and sentenced to two years in military prison. Bugler William Pierce stole a cap valued at $4 from Lee & Reynolds, the sutler's store at Camp Supply in 1870. The court sentenced him to confinement at hard labor under guard for six months and forfeiture of $12 per month for same period. In another case of theft, William Clark stole a pair of pants from James Wilson and sold them to

George Overton of Troop K for $2.75. His sentence of confinement included a 24-pound weight attached to his left leg by a three-foot chain. A punishment that included ball and chain was only to be used in extreme cases, yet it appears that it was frequently used just to make the sentence more severe. In April 1873, Clark got in a dispute at the government ferry for Fort Gibson. The argument escalated and Clark attacked ferryman Charles Reade with an axe. Reade died from his injuries on April 23. For murder Clark was sentenced to be discharged dishonorably and confined to life in prison.[149]

A different Clark from Troop H had also been found guilty of felony offenses. On the night of November 18, 1868, Ellson Clark entered the house of Capt S.B. Lauffer, Quartermaster at Fort Wallace, and stole some jewelry from his kitchen. He then attempted to set fire to the house to cover his theft. Captured and brought to trial, Clark's sentence was "to be indelibly marked with the letter T, one and one-half inches in length, on the left hip," forfeit all pay except that due the laundress, receive a dishonorable discharge, and be confined in the penitentiary for two years. Theft appears to be the only offense for which the convicted felon was branded other than desertion and that punishment had ceased to be acceptable by the end of the frontier period. The 1893 Manual for Courts-Martial stated "No person in the military service shall be punished by flogging, or by branding, marking, or tattooing on the body." Another case where the sentence involved something other than confinement or fine was that of Private William Booter who seemed to have a business on the side dealing in cavalry boots. In October 1868, he sold his issue boots to Henry Wilson of Company I for $1.50. The following week, he stole a pair of boots out of the tent of Saddler James Clayton and sold them to Peter Johnson of Company I for $3. Booter was caught before he could continue his illegal trade and sentenced "to have one side of his head close shaven; to be dishonorably discharged and drummed out of the service."[150]

Some of the cases involving members of Troop H included violent crimes. On May 28, 1875, Blacksmith George Douglas discharged his pistol with intent to injure Andy Clayton. He received a dishonorable discharge and confinement at hard labor for one year. Andy Clayton had been accused of a similar crime in

1874. The prosecution claimed that Clayton had entered the quarters of Lydia Brown, a laundress, and drawing a large knife threatened, "I'll cut you if you don't undress and let me sleep with you." In that case there was insufficient evidence to find Clayton guilty. In 1889, Louis Bell had a woman named Frances Jones in his tent at the San Carlos reservation. They were playing on his bed, when she asked Bell to go to bed with her. As he began to undress, he handed Frances his Schofield Smith & Wesson pistol which had been stuck in his belt while he was packing mules. Somehow, the pistol discharged and Frances Jones fell dead. Bell insisted that he always kept his pistol on an empty cylinder, but it may have shifted while they played on the bed. Bell was also acquitted.[151]

On May 10, 1870, Robert Valentine and Edward Jackson began a heated argument on the way to the stables. Valentine became so angry that he loaded and cocked his carbine yelling at Jackson "that he would fix him." Brandishing his weapon, Valentine managed to shoot a horse by accident. The court reminded him that the only time he should have had a loaded weapon in the stables was on stable guard at night and sentenced him to forfeit $10 per month until the $177 established price of the horse was paid and six months at hard labor. In another disagreement, while in camp at the head of the San Carlos River, Charles Reed attacked Nelson E. Davis and "did seize the left thumb of said Davis in his mouth and bite the same severely to the bone." Reed forfeited $25 and was confined at hard labor until his expiration of service four months later. He was not allowed to re-enlist.[152]

In a quarrel over a card game, Samuel Porter faced down Sergeant William Johnson with his pistol in hand. Leaving the game, he entered a bar where he started another disturbance, drawing his pistol again. Despite his obvious anger, Porter did hand over the pistol when asked. He was confined at hard labor for one year and forfeited $10 per month for same period. In another incident at Fort Davis, Porter himself became the victim of a murderous assault with a knife. John Dupree attacked Samuel Porter, cutting his head four times and attempting to slice him in the side. Dupree claimed that Porter had slapped him and spit on the floor near his bunk. Sergeant Isaac Jackson disarmed and arrested

Dupree. For the attack, Dupree received a dishonorable discharge and six months at hard labor. False accusations could also result in punishment. At Fort Apache, George Horton accused Captain Charles Veile of beating him. When no evidence could be found, Horton was convicted of falsely accusing the captain and sentenced to confinement at hard labor for six months with forfeiture of $10 per month for the same period.[153]

During the same time period, Charles Gray who had been accused of disorderly conduct was further charged with lying in his statement concerning the incident. While not convicted of disorderly conduct, he was sentenced to forfeit $10 for two months for perjury. Another notable case was the incident of the horned toad. Lewis Hayes commanded a detail in charge of grazing the horse herd about a mile from Camp Supply in June 1870. One member of the detail, William Shaw had a horned toad given to him by Pollard Cole and he began showing it to each of the soldiers watching the herd. Hayes carefully avoided the toad, showing some genuine fear of the creature. When Shaw attempted to force the toad on him, Hayes yelled, "Damn you take that frog away!" and "If you don't, I will shoot you." Although Hayes then repeatedly threatened to shoot the toad, the soldiers continued to goad him until he finally stuck his carbine up against Shaw's right hand and fired. By law, no amount of provocation would have been sufficient to justify his wounding Shaw. The court gave Hayes a dishonorable discharge and confined him to a military prison for eighteen months.[154]

This case displays both the comic and tragic sides of a soldier's daily life. As in all civilized society, offenses occurred on a regular basis that needed to be resolved by a military court. Sometimes they were not dealt with in an equitable manner due to circumstances including the difference in composition of each court, the inclination of each individual commanding officer, and the prior history of the accused. Still, the Tenth Cavalry does not appear to have suffered under the military justice system more than any other regiment at the time. The greatest challenge confronting a trooper of the Tenth was simply to survive the hardships, routine and regulations of daily army life. In over thirty years of frontier service, the men of Troop H learned to survive in the harshest conditions in every environment of the western United States. As the first generations of free Black soldiers made a place for themselves in the changing nation, they also carved out a place of honor in the United States Army.

Troop H, Tenth U. S. Cavalry

THE FINAL FRONTIER:
ARIZONA, NORTH DAKOTA AND MONTANA 1885-1898

Goyahkla, better known as Geronimo
Courtesy Oklahoma Historical Society

When Troop H marched out of Fort Davis on April 1, 1885, they joined the other troops of the Tenth Cavalry enroute to the Arizona territory along the route of the Southern Pacific Railroad. For the first time since the regiment's organization in 1866, all twelve troops assembled at Bowie Station just north of Fort Bowie in southeastern Arizona on April 11. From there Troop H received assignment to Fort Grant, Arizona arriving on May 2. Captain Charles Cooper's daughter Birdie described Fort Grant being nestled at the foot of Mount Graham where "From the wide inviting porches of the officers' homes was a view across the Sulphur Spring Valley to the Galiuro Mountains, twenty miles distant." Soon after the company arrived they discovered that Geronimo had again left the San Carlos reservation on May 17 striking out for the Sierra Madre and Mexico. With forty troopers ready for duty, Captain Cooper took the field the next day. For most of the next year, H Troop chased the elusive Chiricahuas through the mountains of southeastern Arizona.[155]

Artist Frederic Remington rode with the buffalo soldiers from Fort Grant on one of the scouting patrols during this period. Preparing to leave the post he recalled, "At the adobe corral the faded coats of the horses were being groomed by black troopers in white frocks." "Marching under the summer sun of Arizona was real suffering and not to be considered by one on pleasure bent," Remington continued. "Uncle Sam's beans, black coffee and the bacon which every old soldier will tell you about would fall to the lot of any one who scouted with the 10th Dragoons." He soon developed an appreciation for the way in which the Black troopers cared for their horses, asserting that "every old soldier knows that his good care will tell when the long forced march comes some day, and when to be put afoot by a poor mount means great danger in Indian warfare. The soldier will steal for his horse, will share his camp bread, and will moisten the horse's nostrils and lips with the precious water in the canteen."[156]

Remington's reminiscences accurately describe life in the field for Troop H in the summer of 1885. The troop soon established a supply camp at Oak Canyon twenty miles south of Fort Bowie in the Chiricahua Mountains scouting for ten to twenty days at a time toward the Mexican border. The soaring rocks and deep canyons of the mountain range made mounted patrols difficult and tracking the Apaches even harder. The fall of 1885, Corporal John Casey remembered, was particularly severe with constant rain, sleet and snow. "We got lost in the mountains for four days being snow bound..." he reported, stating that the patrol had to cut brush and lie on it to avoid the wet frozen ground. Some of the stranded soldiers developed scurvy. Blacksmith Solomon Boller recalled, "That is where I contracted a severe cold that turned into what the doctors called the catarrh of the stomach. I was treated while in the field through snow and rain I had to go. Sometimes snow would be up to our knees and I had to shoe horses in snow and rain..." Randall Blunt of Greenville, North Carolina who developed rheumatism on these patrols asserted "there was much snow in the mountains, the weather was very stormy, in chasing and capturing Chief Geronimo and his bunch." During January the troop moved their scouting camp to Bonita Canyon, in the Chiricahuas southeast of the fort, where the springs provided a reliable source of water. An added benefit in Bonita Canyon was the plentiful fruits and vegetables provided by local farmer J. H. Stafford. There on the floor of a rugged, rock strewn canyon, they established a semi-permanent outpost with Troop E. While in camp, the soldiers built a rock monument to honor assassinated President James Garfield.[157]

Garfield monument built by Tenth Cavalry soldiers in Bonita Canyon.

Major Louis Carpenter commenting on his old troop's campaign from his current post in Nebraska cautioned, "The character of the country is such that a half dozen Indians can burn, plunder and massacre in the face of a couple of companies of troops and escape to the mountains or dodge across the line into Mexico before much can be done towards their capture." In January 1886, Lieutenant William Shipp of Troop H, described by Birdie Cooper as "a tall, bashful chap, very easy embarrassed," accompanied a party of Apache scouts under Captain Emmet Crawford into Mexico to parley with Geronimo. Before talks could begin, a Mexican militia force attacked Crawford's Apache scouts, believing them to be Geronimo's band, and Captain Crawford was killed. Still desiring to make peace, Geronimo agreed to a conference with General George Crook at Canyon de los Embudos in the Sierra Madre on March 27, 1886. At that conference, the Apaches agreed to surrender and go east for two years before returning to Arizona. Crook agreed to the Apache terms, but General Sheridan disagreed, insisting on unconditional surrender.

Sheridan's displeasure caused Crook to submit his resignation. He was replaced by General Nelson Miles. While the generals debated, Geronimo and Nachez escaped with twenty men and sixteen women and children into the mountains. General Miles ordered all of the Apaches rounded up and deported to Florida. Over the next few months, Troop H and the rest of the Tenth Cavalry were tasked with chasing down renegades and preparing the reservation Indians for transport. Even some of the Apache scouts were deported. When Geronimo discovered that there would be no Apaches left at San Carlos, he agreed to surrender to General Miles in person. On September 4, Geronimo laid down his arms at Skeleton Canyon and departed for Florida, never to return to his homeland.[158]

With Geronimo gone, only one Apache leader remained at large. Mangus, son of Mangus Coloradas, still led a small family band in the mountains south of the Mexican border. In late September, Lieutenant Carter P. Johnson of Troop M followed Mangus north out of the Chihuahuas onto a ranch where the Apaches stole a herd of mules and then moved on through the Black Range and Mongollon mountains. On October 14, Colonel James F. Wade sent Captain Cooper with a patrol of twenty men and two Indian scouts from Fort Apache to find Mangus. The Troop H detachment consisted of Sergeant Pollard Cole, Sergeant Isaac Jackson, Corporal John Casey, Corporal William Hawkins, Trumpeter Silas Jones, Blacksmith Solomon Boller, Walter Armstrong, William P. Battle, Joseph Cammel, James Dillard, James Gibson, Charles A. Green, Colonel E. Miller, Joseph Rousey, Charles L. Terry, George W. Foster, Thomas Bruff, Gabe McKibbin, George W. Newman, and Augustus Sparks.[159]

Proceeding northeast from their old camp on the Bonito Fork, they found the Apache trail leading into the White Mountains on October 17. Cooper reported, "I took up the trail and followed it at as rapid a gait as possible, considering the kind of country I had to travel over, it being almost impossible to describe its ruggedness." After a forced march of thirty miles they finally sighted the Indians going over the summit of a mountain almost 2000 feet high. The troopers took up the chase climbing their horses over five mountain peaks and fifteen miles of "rugged and almost inaccessible country." According to Corporal Casey, "The mountain was so steep we had to hold the horses back to keep them from falling down." Casey and Sergeant Cole scouted ahead. In Casey's recollections, he states that six soldiers got ahead of the rest of the detachment because "the Captain being a very large man had to stop and rest very often." Those six were the first to discover that the Apaches had abandoned their stock and concealed themselves in a T-shaped canyon.[160]

The soldiers fired shots from both sides of the canyon to encourage the Indians to surrender. All but Mangus and two others did surrender or were captured while trying to hide from the soldiers. The troopers set up camp for the evening, preparing supper for themselves and their captives. Captain Cooper reported, "I sent one of the squaws out into the mountains to induce the rest to come in." Using Casey to translate the Apaches' Spanish, Cooper negotiated with Chief Mangus through the woman captive. Finally at about 8:00 a.m. the following morning Mangus agreed to surrender. Birdie Cooper later claimed that her father told her Mangus declared "I am not ashamed to surrender to you. You are a better soldier than I am." When negotiations ended, Cooper's detachment had captured Mangus, two warriors, three women, "two boys able to bear arms," one girl, four children, twenty-nine mules and five ponies. The two older boys were the sons of Victorio and Juh who had come under the care of Mangus and one of the other children was his own son, Frank Mangus. According to a story written by Birdie Cooper, Mangus rode side by side with Captain Cooper as they entered Fort Apache.[161]

The capture of Mangus proved to be the last action Troop H would see in the Indian Wars. Members of the detachment received honorable mention for their part in apprehending the last Apache chieftain. John Casey was returned to his previous rank of sergeant. William P. Battle received promotion to corporal. Some of the troopers involved in the capture received what they considered a special honor. They guarded the Apache captives on their long train ride across the southern United States to Saint Augustine, Florida. The honor of the situation seemed a little dubious when on November 3, about three miles east of Pueblo, Colorado, Mangus jumped through a window on the moving train in an attempt to escape. After stopping the train, troopers found him bleeding and unconscious on the tracks. Back on board, Mangus slipped one arm out of his handcuffs and used them for a weapon, knocking Sergeant Casey to the floor and fighting off the other soldiers from his concealed position under the seats. Finally Mangus was dragged out and while the other prisoners were held back with cocked rifles, he was bound with bell cord until the troopers could manacle him. After being subdued, Mangus spent the rest of the trip in manacles and leg irons. Back in southeastern Arizona, regular patrols continued to cross the rugged mountain terrain. Occasionally, soldiers would still find the trail of scattered renegades from the San Carlos reservation, but after more than twenty-five years of fighting the Apache wars had come to an end.[162]

Apache captives enroute to Florida in 1886.

Now stationed at Fort Apache, H Troop continued to perform regular drill and fatigue duties. In 1887, Private Colonel Miller, a member of the expedition against Mangus, suffered a serious accident while at rifle practice. "I got struck in eye by a cartridge shell in May 1887," he wrote, "Was ordered by Sergeant to blow rag out of the gun, and as I snapped it and it would not go off and I sprung the chamber and cartridge shell flew out and struck my eye." Even without Indian hostilities, accidents and illnesses continued to take a toll on the troop. On December 15, 1888,

Sergeant Robert B. Banks died of exhaustion, gastritis and old age. Banks, who served for many years as a farrier, had joined Troop H during its formation more than twenty years before in 1867.[163]

After twenty-two years with the Tenth Cavalry, Colonel Benjamin Grierson left to assume command of the Department of Arizona in November 1888. He penned these words in praise of the regiment he had commanded since its organization, proudly asserting they were "always in the vanguard of civilization and in contact with the most warlike and savage Indians of the plains. The officers and enlisted men have cheerfully endured many hardships and privations...and they may well be proud of the record made..." Colonel John K. Mizner, a Civil War veteran who had served with the Fourth and Eighth Cavalry on the frontier, succeeded Grierson as commander of the Buffalo Soldiers of the Tenth.[164]

In 1890, Michael Finnegan, a sharpshooter who had joined Troop H in Texas, became its First Sergeant. Charles Gray and Silas Jones were trumpeters, Louis Bell, blacksmith, John Thompson, saddler, and Lewis Harris, farrier. Blacksmith Louis Bell deserted from Fort Apache on June 16, 1890. He was captured five miles from post the following morning, sentenced to confinement under guard at hard labor for two years and given a dishonorable discharge. That year ended with the last blow of the Indian Wars far to the north at Wounded Knee Creek on the Pine Ridge agency in South Dakota. On December 29, soldiers of the Seventh Cavalry surrounded the Indian encampment in an attempt to disarm Big Foot's Sioux. One of the Sioux accidentally fired a shot and in the ensuing confusion over 200 Indian men, women and children died.[165]

The 1890s also brought many changes nationally as the use of electric power and telephones continued to spread and automobiles began to appear on the scene. Sequoia and Yosemite joined Yellowstone as officially protected national parks in the autumn of 1890. Thomas Edison opened his first motion picture studio in 1893. For H Troop the first major change came in February 1891 when Captain Cooper transferred to the command of Troop M. For the first time, the officer who took command of Troop H had not served during the Civil War. He was Captain Thaddeus W. Jones, a graduate of the U. S. Military Academy in 1872. Jones, born in Henderson County, North Carolina on July

31, 1848, married Mary Lee of Michigan and four children were born to them while he served on the Indian frontier. As first lieutenant, Jones commanded Company B, Tenth Cavalry during the campaign against Victorio. His company served beside Troop H at Rattlesnake Springs, in pursuit of Victorio, and in patrolling the Rio Grande from Ojo Caliente. He served as regimental adjutant at Fort Grant in 1890. Jones received his promotion to captain on January 14, 1891 and assumed command of his new troop the following month. The next change for Troop H occurred in April 1892, when the Tenth Cavalry moved its base of operations from the southwest to Montana and North Dakota. Noting that the Tenth had been based in the south for over twenty years, Colonel Mizner had requested an assignment north of the thirty-sixth parallel, probably having the plains of Kansas in mind. Instead the Tenth was bound for the far north. Troop H left Fort Apache on April 21 by rail arriving at Fort Buford, North Dakota on May 5 in the middle of a blizzard.[166]

Fort Buford, founded in 1866, stood at the confluence of the Missouri and Yellowstone Rivers in the northwestern corner of North Dakota. Troop H soon took the field again patrolling over the flat rolling lands of western North Dakota and eastern Montana. In September 1892, they participated in a practice march of 114 miles. Even in the relative peace of the northern frontier, the hardships of army life continued. Private James Willard died on July 1, 1893, drowned in the Missouri River near the post. July 1894, found the troop in Montana guarding the property of the Northern Pacific Railroad against strikers. For the first time, the federal government had been granted an injunction against striking workers in the Pullman Strike which had been ongoing since May. The strike had spread to twenty-seven states threatening to bring passenger service on the railroads to a stop since every passenger line used Pullman cars. President Cleveland called out troops under General Nelson Miles to break the strike at the Pullman works in Chicago and called for troops throughout the country to protect the trains.[167]

Railroad and bank failures the previous year had put many out of work and caused labor conditions to deteriorate across the nation. What became known as the Panic of 1893 began in June with a stampede of stock sales leading to a market crash. With the failure of over 150 rail lines and 500 banks, railroad strikes became increasingly common, culminating with the strike against the Pullman company. Over the next four years unemployment grew to over ten percent with one out of every six workers losing a job. As had happened in 1873, the depression caused a marked increase in enlistments between 1893 and 1896. The number of enlistments in H Troop during that period nearly equaled the number who joined when war broke out with Spain in 1898. For one of the few times in its history, the troop maintained a full roster with some months bringing ten to fifteen new recruits. During a period of great economic uncertainty, the army continued to provide a positive opportunity for young Black men.[168]

When the H Troop marched out to guard the railroads, Captain Jones did not accompany them. After only three years, he had been transferred. His replacement, Captain Levi P. Hunt had been with the Tenth Cavalry since 1870. Hunt was born August 7, 1845 in Bowling Green, Missouri. He attended the U. S. Military Academy graduating fifty-eighth in the class of 1870. He was first assigned as second lieutenant of Company A, Tenth Cavalry at the same time that Charles Cooper served as first lieutenant. After transferring to Company E at Fort Concho, Hunt met and married Mahala Badger, daughter of Chaplain Norman Badger. The chaplain had become fatally ill while stationed at Fort Concho. Mahala, known to family and friends as Haidie, apparently suffered from some of the same ailments. Hunt's company was stationed at Fort Quitman, patrolling the Rio Grande around the time of the Victorio campaign when Haidie died. Lieutenant Hunt married a second time at Fort Davis to Susan Murphy, daughter of a prominent ranching family. Levi and Susan had two children, Claude de Bussy and Ellen Louise, known as Nell. The Hunts became good friends of the Coopers, beginning with the two lieutenants' service together in Company A. Colonel Mizner described Hunt as "a good average everyday man." In July 1894, at the age of forty-eight, he took on the command of H Troop.[169]

Sergeant Pollard Cole retired on August 13, 1894 after twenty-seven years of service. He had been at the rescue of Forsyth's scouts, the defeat of Victorio at Rattlesnake Springs and the capture of Mangus. Cole was one of the first to retire from the many old soldiers who had spent their entire career in H Troop.

William P. Battle, another member of the patrol that captured Mangus, left the service in 1894 after his second enlistment. He married and took a job with the Bureau of Mines in Pittsburgh, Pennsylvania. Other soldiers chose a less conventional way to leave the service. For many years desertions had been only a minor problem in Troop H. In 1893 and 1894 the troop had no deserters. Perhaps because of continuing isolated conditions and severe weather, 1895 proved to be a particularly bad year. From March through August, the troop had one soldier desert almost every month. Two of them, Abe Chambers and Albert Scott had only enlisted in the Tenth Cavalry a few months before deserting. Due to a steady level of enlistment during this period, there were still twice as many soldiers joining the company as those who deserted.[170]

On October 1, 1895, the troopers left Fort Buford traveling on the Great Northern Railway to their new post at Fort Assiniboine, Montana. Fort Assiniboine, built in 1879 to guard the border and prevent Sitting Bull from returning from Canada, was the largest fort west of the Mississippi. Located in central Montana thirty-eight miles from the Canadian border, it served as headquarters for the Tenth Cavalry during the 1890's. H Troop also received a new assignment of horses. During the 1890's, they could no longer claim to be the black horse troop. Instead, they received a mixed allotment of sorrels, greys and roans.[171]

The year 1897 brought more changes for the men of Troop H. In June Colonel Mizner received promotion to brigadier general and in June resigned command of the Tenth Cavalry. Colonel Guy V. Henry replaced Mizner. In his last order, Mizner praised the soldiers' "devotion to every duty, however trying or arduous. For efficiency and discipline and valuable service the Regiment has a record of which it may justly be proud." James H. Thomas, who had joined Company H back in 1867, was appointed Chief Musician of the Tenth on July 7. Thomas had begun his career as a musician in the First USC Infantry during the Civil War. He transferred from Company H to the regimental band in 1876 and became chief trumpeter in 1891. He retired after a distinguished career on August 3, 1897.[172]

Thomas was only one of the old soldiers of H Troop to retire after thirty years of service on the frontier. Trumpeter Silas

Jones also retired on July 7. The final personnel report on Jones listed $44.23 retained pay, $28.63 clothing not drawn and rated his character as "Excellent." Saddler Joseph Claggett of Maryland followed him into retirement on September 20. Claggett also received a character rating of "Excellent," leaving with $43.08 retained pay and $17.83 clothing not drawn. The two old comrades settled down in the town of Havre, Montana just outside of Fort Assiniboine. After sounding the bugle in dozens of engagements over thirty years and thousands of miles with Troop H, Silas Jones married and with his wife Nellie continued to live in Havre well into the twentieth century.[173]

Inspector General Colonel George H. Burton inspected the camp of Troop H on September 3 - 16, 1897. As part of their readiness inspection, they performed a practice attack and defense of the camp. On September 20, they mounted a practice march to Lewiston where they engaged in another practice attack and defense. All of the drill and practice must have seemed very different to some of the old soldiers who had spent years chasing Indians across the southern plains, riding and camping in all types of terrain and weather. Frederic Remington, riding along with Troop H during a march the previous year had commented, "Some of the old sergeants have been taught their battle tactics in a school where the fellows who are not quick at learning are dead."[174]

As Remington rode near the rear of the column with Shelvin Shropshire who had been First Sergeant of Troop H since 1891, Shropshire regaled the artist with stories of the old days. The first sergeant concluded by saying, "Ah, Mr. Remington, we used to have soldiers in them days. Now you take them young fellers ahead thar - lots of 'em 'il nevah make soldiers in God's world. Now you see that black feller just turnin' his head; well, he's a 'cruit, and he thinks I been abusin' him for a long time. Other day he comes to me and says he don't want no more trouble; says I can get along with him from now on. Says I to that 'cruit, 'Blame yer eyes, I don't have to get along wid you; you have to get along wid me. Understand?'" Remington had showed his respect for Shropshire and the other soldiers of the Tenth in an earlier story when he commented, "Officers have often confessed to me that when they are on long and monotonous field service and are troubled with a depression of spirits, they have only to go about the campfires of

114

the negro soldier in order to be amused and cheered by the clever absurdities of the men." "As to their bravery, I am often asked, 'Will they fight?' That is easily answered. They have fought many, many times."[175]

At the end of 1897, most of the fighting appeared to have ended for the men of Troop H, Tenth Cavalry. After thirty years of active service, many of the original members of the troop had been discharged or retired. Many of the current soldiers had joined after the end of the Indian campaigns. Yet they soon would have the opportunity to see the value of endless hours of practice marches and drills. In 1898, some of them would be called to action on new campaign fronts and new battlegrounds outside the boundaries of the United States.

INTO THE BREECH:
CUBA AND THE PHILIPPINES 1898-1909

Medal of Honor recipient Dennis Bell
Courtesy Library of Congress.

In January 1898, President William McKinley ordered the battleship *Maine* to anchor in Havana harbor in order to protect American citizens in Cuba. A newspaper for the Black community, the *Richmond Planet*, reported the *Maine* dropping anchor on January 26, commenting "The arrival of the warship caused much surprise and excited considerable curiosity." Diplomatic exchanges were completed peacefully with Spanish and German ships, yet Spain was not pleased with this interference and Madrid newspapers talked of sending a Spanish fleet to visit American ports. Many American businesses were strongly against war with Spain. However, the media, especially the presses of Joseph Pulitzer and William Randolph Hearst in a continuing battle for circulation sympathized with the Cubans and tried to create an atmosphere favorable to joining their revolution. They presented the Cuban nationalists as patriots engaged in guerilla warfare against an oppressive governor, Valeriano Weyler. Hearst, replying to artist Frederic Remington's comments that there did not seem to be much of a war in Cuba insisted, "You furnish the pictures and I'll furnish the war."[176]

On the night of February 15 a violent explosion occurred in the bow of the *Maine*. Captain Charles D. Sigsbee reported, "*Maine* blown up in Havana harbor at 9:40 and destroyed. Many are wounded and doubtless more killed and drowned. The wounded and others are on board the Spanish man-of-war *Alfonso XIII* and Ward line steamer." Despite strained relations between Spain and the United States, the Spanish cruiser *Alfonso XIII* sent out all of its boats to rescue crewmen from the *Maine*. The ship could not be saved and 260 of the crew were lost. Ironically, the *Alfonso XIII* was later sunk trying to run an American blockade of Havana. Although there was no evidence that the Spanish had any part in the destruction of the *Maine*, the press declared it another Spanish outrage and support for the cause of Cuban independence increased throughout the United States.[177]

Far to the north at Fort Assiniboine, Montana, drill and fatigue duties continued to fill life on post for the men of Troop H. In January 1898, Lucilius Drane and James H. Alexander received promotion to sergeant. On February 8, Sergeant Charles Faulkner of H Troop was named national standard bearer for the Tenth Cavalry. Faulkner was a veteran, having first enlisted at Fort Davis in 1879. On his next enlistment in 1902 he would write, "I have the honor to request that I be allowed to re-enlist in Troop H 10th U.S. Cavalry as a married man (no children), character 'Excellent' and am serving in the 23rd year of continuous service."[178]

Sergeant Faulkner did receive permission to re-enlist; however, that was not the case for all married men. General Order 5 of March 12, 1898 stated that all enlisted men wanting to marry must submit applications to the regimental commander "for his approval or disapproval; the latter will generally follow, unless the man has had over fifteen years service in the Army with other cogent reasons." The order also declared that there would be no quarters furnished for enlisted men who married and those who married without their commanding officer's permission would not be allowed to re-enlist. General Order 6 concerning married enlisted men confirmed, "It is thought that the limit of two to each troop or organization should not be exceeded..." Troop H did not conform to this regulation in 1898. They had seven married enlisted men.[179]

Although marriage may have been one of the more important peacetime issues for the Tenth Cavalry, even in remote Montana, they followed the news from Cuba and awaited their marching orders as professional soldiers. President McKinley recognized Cuba as an independent nation and Congress declared war on Spain on April 21, 1898. Sergeant Major Edward Baker recalled that even before that date the Tenth had received a telegram on April 16 ordering their transfer to the Department of the Gulf in preparation for hostilities. On April 19, the regiment boarded trains leaving only a maintenance detachment behind at Fort Assiniboine. As they rode the rails south, Baker remembered that at first they were "heartily greeted" and "every little hamlet, even had its offerings." When the troop train wound its way into the southern states, the cheering died out and signs of segregation became more prevalent. The Supreme Court case, Plessy vs. Ferguson, had upheld the legality of "separate but equal" facilities as an acceptable solution in 1896. Throughout the 1890s southern states continued to pass laws enforcing segregation of public facilities, public schools and public transportation. After years of service on the open plains and unsettled regions of the west, the Tenth Cavalry had now entered the "Jim Crow" south.[180]

On April 25, they arrived at the recently established Chickamauga and Chattanooga National Military Park in northern Georgia. Proximity to the major railway hub in Chattanooga and an expanse of open land made the park an excellent site for marshalling an army. Here the regular regiments of the United States Army assembled and drilled together before proceeding to Florida to board transports for Cuba. Chaplain Theophilus G. Steward described Chickamauga as "beautiful by nature, especially in the full season of spring when the black soldiers arrived there, and adorned also by art." First Sergeant Peter McCown of Troop E recalled the more practical aspects of camp, "We were drilled every day in battle exercises." The regulars only stayed at Chickamauga a brief time. By May 14 they were on the way to Florida by various routes and volunteer regiments were mustering on the same campgrounds at Chickamauga. Due to lack of usable water for the troops at Tampa and the difficulty of transporting large numbers at once on a single-track railroad, the Tenth Cavalry camped at Lakeland, thirty miles from the Florida coast.[181]

At Lakeland, the white citizens were furious at the prospect of having Black troops quartered in their quiet little town. They refused to distinguish between Black soldiers and Black civilians, insisting that the troopers observe strict racial segregation. Several incidents occurred before the Tenth Cavalry could leave for Tampa on June 7, 1898. A white barber died while using his pistols to keep two Black soldiers from being served in a local drug store. Added to the problems of racial discrimination, the poor sanitary conditions in Lakeland led to an outbreak of typhoid fever. Unfortunately for the troopers, not all of the regiment left Lakeland. The Tenth had been divided into three squadrons and the Third Squadron was detailed to remain behind with the horses and extra supplies. For a change, H Troop did not march in the vanguard of the Tenth Cavalry. Assigned to the Third Squadron, they remained in Florida under the command of Captain Levi Hunt.[182]

While most of Troop H remained at Lakeland, some of the company's former members did play an active role in the Cuban campaign. William Battle, whose enlistment had ended in 1894, was a member of Battery B, Pennsylvania Light Artillery in Philadelphia when the war began. He immediately re-enlisted in the Tenth and later claimed to be the first bugler to sound the charge at San Juan Hill. Thomas Bruff, a member of the patrol that captured Mangus, went to Cuba with the regimental band, but did not return. He became one of the many victims of yellow fever. Former H Troop lieutenants Charles Ayres and William Shipp both embarked with the Tenth from Tampa on June 14. Charles Cooper mustered volunteers in New Mexico to join the First Volunteer Cavalry, known as the Rough Riders because of a comment to the press by Theodore Roosevelt concerning the type of men he wished to recruit. Their commander, Colonel Leonard Wood praised Cooper's efforts stating, "He has recently brought the New Mexico quota of my regiment to the rendezvous here in excellent condition and organization." Theodore Roosevelt who said the "difficulty was not in selecting, but rejecting men," later recalled, "Colonel Cooper rendered very great aid to me when I was beginning my duties as Lieutenant Colonel of the First Volunteer Cavalry, both in enlisting men and in helping to equip us at San Antonio. I was greatly struck by his knowledge and efficiency." One thousand men and 1200 horses and mules left San Antonio on May 29 for Lakeland, Florida.[183]

Colonel Louis H. Carpenter (seated) with his staff.

The beginning of 1898 found Colonel Louis Carpenter in command of the Fifth Cavalry as well as the post of Fort Sam Houston in San Antonio, Texas. Early in the year, he had taken the Fifth on a practice march accompanied by one of his former Troop H lieutenants, Alexander Keyes, now medically retired. When the war began, Carpenter organized and carried out the mobilization of the Eighteenth Infantry and the Fifth Cavalry, establishing a line of supply and insuring the care of military dependents. He accompanied the Fifth as far as New Orleans, at which point he received orders assigning him to the command of volunteers mustering at Chickamauga Park. On May 4, 1898, Carpenter received an appointment as Brigadier General of Volunteers. He transported a brigade of volunteers to Tampa on May 30 and reported to General William Shafter. Carpenter's brigade encamped on the Tampa River where he claimed "there was some shade and fine bathing for the men, although the space was somewhat contracted." Heavy summer rains soon led to typhoid and unsanitary conditions. Carpenter moved his camp to

Fernandina, Florida, where he "was forced to make large purchases in the open market of forage and of medical material and supplies, the animals not having been provided for and the number of sick in hospital being greatly in excess of those estimated for." Although his volunteers remained ready for action, they never did embark for Cuba. By the early part of July, the fighting had ended. On August 11, Carpenter moved his troops to the newly established camp at Huntsville, Alabama.[184]

Although Colonel Carpenter and most of Troop H remained in Florida, the first two squadrons of the Tenth Cavalry under the command of Lieutenant Colonel Theodore A. Baldwin boarded the transport ships for Cuba on June 14, 1898. They landed at Daquiri, a shipping port of the Spanish-American Iron Company on June 22. On the 23rd, the dismounted cavalry marched inland through heavy rains. Frederic Remington, once again traveling with the Tenth, complained about the wastefulness of dismounting cavalry and commented, "I am as yet unable to decide whether sleeping in a mud-puddle, the confinement of a troop-ship, or being shot at is the worst." Traveling on empty stomachs, the troops marched until 10:00 p.m. and then began their advance again at 5:15 the next morning. Led by Brigadier General Joseph Wheeler, a former Confederate cavalryman, the dismounted cavalry clashed with the Spanish rear guard around 7:45 a.m. at Las Guasimas.[185]

At first many of the Tenth Cavalry soldiers remained in an awkward position in the underbrush, with many of the First Volunteer Cavalry in front of them and no clear line of fire. Captain Charles Ayres recalled, "The troop stood for an hour and a half under a terrible fire without firing a shot in return for fear of killing our own men in front, who were in the bushes." Lieutenant Colonel Roosevelt of the First Volunteers and Ayres helped organize the line and gradually increased their fire against the Spanish. At one point Ayres exposed himself to heavy fire, when he and four troopers carried Major James M. Bell of the First Cavalry, whose leg had been shattered, to safety behind the lines. The Americans increased their fire, using a battery of Hotchkiss mountain guns, until the Spanish line finally broke. General Wheeler, observing their retreat, reputedly yelled, "We've got the damn Yankees on the run." Lieutenant William Shipp, serving in a temporary assignment as brigade quartermaster, came up to the lines that evening to congratulate his comrades on their victory.[186]

General Shafter determined to rest and regroup while his supplies were brought up, before launching a frontal assault on San Juan Heights where Spanish troops guarded the approaches to Santiago. On the evening of June 30, the cavalry division stood ready at El Pozo just across the San Juan River from the heights. The next morning, Captain Ayres wrote "At daylight when the sky was crimson as the earth was soon to be" the artillery opened fire and after about an hour "the cavalry division swept like a flood down from the top of the El Pozo position up the road to Santiago." Rough terrain, tropical heat and heavy fire while crossing the stream all slowed their advance. Lieutenant Shipp brought word for the Tenth to form on the left of the First Volunteer Cavalry. Some of the officers teased him about riding back and forth with messages, to which Shipp replied that he would really rather be with his troop.[187]

It was almost noon before the troopers were ready for the final assault. Using the Hotchkiss guns to cover their advance, cavalry and infantry divisions charged San Juan Heights. Initially the Tenth was ordered to provide covering fire for the Rough Riders. When the volunteers stalled under heavy fire, the Black troopers moved forward cutting through barbed wire and opening a path for the cavalry to advance. The cavalry division led by the Rough Riders took the summit of Kettle Hill to the right of San Juan Hill "firing steady and deadly volleys, with the enemy's shells screeching and bursting over their heads." After leaving Private Henry Jackson in charge of the mules, Lieutenant Shipp joined the advance on the heights, trying to rejoin his troop. Soldiers found his body later, partway up the hill, shot through the heart. As the soldiers crested the heights, the Spanish gave way, retreating to a second line of blockhouses and defensive works. By 1:30 p.m. Americans held San Juan Heights. The day had been very hot and many canteens had been pierced by bullets. Remington wrote, "Our men sat about in little bunches in the pea-green guinea-grass exhausted." The cavalry division soon dug in and began to prepare their defenses. That night Captain Ayres assisted Colonel Roosevelt and his men digging rifle trenches for their regiment.[188]

For several days, the Tenth remained in position, exchanging fire with the Spanish defenses. Then on July 4, a truce was declared. The Richmond *Planet* reported, "Today is a glorious Fourth for all races of people in this great Land." Sergeant David T. Brown of Troop H, working in the recruiting office in Atlanta exclaimed, "It made my heart swell with joy and pride when I learned that my regiment had taken such a noble stand..." Troop H Corporal John E. Lewis wrote, "About every troop of the 10th lost its officers ... and non-commissioned officers took their places and led the troops on to a victory that has gained the admiration of the world." "The troops had confidence in their Negro leaders: they did not become demoralized, but marched on to a glorious victory under the leadership of Negroes whose names should go down in history." On July 10, naval and land forces began a bombardment of Santiago. On July 17, the Spanish at Santiago signed terms of surrender.[189]

While the first two squadrons of the Tenth Cavalry fought on the front lines around Santiago, selected members of the Third Squadron engaged in an operation to reinforce and resupply Cuban insurgents. Lieutenant Carter P. Johnson of Troop H, at one time the commander of Apache scouts for the Tenth in the search for Geronimo and Mangus, led the special detachment. He chose fifty soldiers from the troops still camped at Lakeland. Charles S. Allen, William Harris, Thomas Mitchem, Dennis Bell, Frank C. Henry, William K. Porter, Nathaniel Bullock, Lelwood Loving, and John S. Williams of Troop H were chosen for Johnson's expedition. The detachment left Lakeland on June 21, 1898 and boarded their ships four days later on June 25. Their convoy included the transports *Florida* and *Funita*, escorted by the gunboat *Peoria*. On board were Johnson and his fifty men, Brigadier General Emilio Nunez with about 375 Cuban soldiers, and a sixty-five mule pack train of supplies. Crowded aboard the transports, the troopers had to place a special watch on the horses and mules which were stabled near the boiler room where the heat was always intense.[190]

On June 29, General Nunez chose a landing spot for his troops at the mouth of the San Juan River. Unfortunately, heavy gunfire from the Spanish batteries kept the landing party from making it to shore. The following day, they tried again, this time at the mouth of the Tayabacao River. Landing boats discharged over

300 soldiers, but they could not establish a position and were forced back to the boats. Luck continued to fail them as they attempted to evacuate. Two of the small boats had been sunk and several of the wounded including American volunteer, Winthrop Chandler, had to be left ashore. As the boats pulled away from shore, those unfortunate few lay suffering in the surf, vainly awaiting a rescue attempt. Back on the transports, Lieutenant Johnson worried about the possible capture of those men left on the beach especially since none of them were uniformed soldiers. Four times, he tried to put boats ashore, while the *Peoria* continued to provide a covering fire to hold the Spanish back.[191]

Finally Johnson asked for volunteers from the Tenth and chose five: Lieutenant George P. Ahearn, Sergeant William H. Thompkins of Troop G, Private Dennis Bell of Troop H, and Corporal George Wanton and Private Fitz Lee of Troop M. The five troopers rowed ashore by full moonlight. Slowly they worked their way along between jungle and surf searching for the wounded men. As Spanish gunfire increased, the American, Chandler, hailed the rescuers, guiding them to those stranded along the shore. One by one, they pulled the wounded off the beach and into the boat as Spanish bullets splashed all about them. As soon as they reached the transports, Johnson put out to sea. While the gunboats *Peoria* and the newly arrived *Helena* continued fire, attempting to make the Spanish think the landing was still in progress, the transports steamed forty miles east and put the Cubans ashore at Palo Alto. For a week, Lieutenant Johnson unloaded his cargo without any further threat of attack. He returned to Florida with the idea of mounting another expedition, but the war ended before his plans could be put into action. Johnson recommended all four enlisted men involved in the rescue at Tayabacao for the Medal of Honor stating that , "...the rescue was pronounced by all who witnessed it, as a brave and gallant deed, and deserving of reward.". On June 30, 1899, Captain Hunt presented Dennis Bell, who had enlisted with his brother Arthur in 1896, the first Medal of Honor ever awarded to an enlisted member of Troop H.[192]

The troops returning from Cuba were routed through Camp Wikoff on Long Island, New York where many of them were treated for malaria, typhoid and yellow fever. Corporal Robert Anderson of Troop H died of yellow fever at Siboney, Cuba in

August. Charles K. Terry, who was only a recruit when he participated in the capture of Mangus, had gone to Cuba with the Tenth Cavalry band. Wounded in the side during the assault on Santiago, he returned to his home in Indianapolis to recuperate. Captain Ayres returned to the United States amid great acclaim for his cool and capable leadership. Theodore Roosevelt wrote, "[I] was an eye-witness of the way he handled his men in camp, on the march, and in battle. I saw repeated instances of his coolness, his energy and his conspicuous gallantry." The old Confederate, General Wheeler introduced Ayres before the Tenth Cavalry in Washington, DC claiming "his conspicuous gallantry and highly distinguished services won the admiration of all who saw him." To the Tenth and the other regiments of his division, Wheeler gave a touching tribute, "Wherever my steps may lead, my heart will always burn with increasing admiration for your courage in action, your fortitude under privation and your constant devotion to duty in its highest sense, whether in battle, in bivouac or upon the march." Ayres replied to the praise he received with his own tribute stating "he wished honor and credit given to the men in the ranks who carried the carbines, who dug the trenches, stood in them, and filled them with their dead bodies, covering up their comrades."[193]

Although hailed as heroes in New York and Washington, the Tenth Cavalry still had some of its hardest fighting ahead when the troopers returned to the racial tension of the southern United States. The reunited Tenth arrived at their new station, Camp Albert G. Forse in Huntsville, Alabama, on October 11, 1898 to a volley of bullets fired at their railway cars. The anger of local citizens at having Black troops stationed in their community proved even worse than it had been in Lakeland. Concerning their future deployment, Corporal John E. Lewis of Troop H, writing to the Richmond *Planet* protested, "It seems to be our fate (in spite of the service we have rendered to the U. S.) to return to the dreaded and unhealthy climate of Cuba. But! despite that fact we are willing to go anywhere in preference to staying here." "Why did the government disarm us and send us here among these race-prejudiced whites?" To add to the troopers' discomfort, Black owned newspapers, like the *Planet*, reported lynchings of mostly Black men every week for everything from murder and rape to "struck a white man" or "writing insulting letter." Many of these lynchings took place in Georgia, Florida, and Alabama, where H Troop had been stationed.[194]

On the evening of November 11, soldiers of Troop H experienced what appears to be a further incident of blatant racial hatred. With racial tension so evident, few passes had been granted for soldiers to leave camp. On the 11th, Troop Clerk John R. Brooks and Corporal Daniel Garrett left camp to go into town without receiving permission from the officer on duty. A local Black citizen known as "Horse" Douglass followed them. Douglass ambushed the pair and shot them as they were returning to camp. He killed Brooks on the spot and mortally wounded Garrett. When questioned, Douglass claimed to be in the pay of a conspiracy of white citizens intent on murdering members of the Tenth Cavalry as long as they remained in Huntsville. The conspiracy would hire Black men to perform the actual murders, promising them protection from prosecution. While there appears to have been no proof of this plot, within a month Douglass had been set free and was observed in a local saloon "flourishing a revolver and making threats" against all members of the Tenth Cavalry.[195]

The Richmond *Planet* mourned the death of John Brooks, who had worked at the newspaper from the time he was thirteen until he enlisted in April 1896. He was only twenty-one when he died. Brooks left behind his father, mother, two brothers and three sisters. His funeral was held on Sunday, November 20 at Ebeneezer Baptist Church in Richmond, Virginia. Corporal Garrett survived until November 14, but there was little that could be done for him. He was buried at the camp in Huntsville. John Lewis protested, "For God's sake if not for ours, why do they not remove us to some place, no matter where, so long as we are where one might leave the camp at 2 p.m. and go in town without being afraid of being murdered." He continued, "Take us away from Alabama and send us to Cuba or even Hades."[196]

With morale understandably low, Captain Hunt decided to use all of his resources to plan a first-class Thanksgiving dinner for the troop. Families were invited to join their soldiers from as far away as Fort Assiniboine and Washington, D.C. Troopers set up two thirty-foot tables to be filled with food. Chief Cook Mack Harris and Cook John White assisted by Frank Boyle and John

Wallace prepared roasted turkey stuffed with oysters, cranberry sauce, roast pork, opossum baked with sweet potatoes, mashed potatoes, baked macaroni, sugar corn, canned peaches, assorted cakes, tea and coffee. The new troop clerk, John E. Westfall of New York City served as master of ceremonies. First Sergeant Shropshire missed the festivities because of a sore foot. Troopers brought him food from the feast and served him in his tent.[197]

In January 1899, Troop H finally left Alabama for Fort Sam Houston in Texas in preparation for their return to Cuba. Along the way to San Antonio the feeling against Black soldiers was so strong at times that the railroad cars were fired upon in Meridian, Mississippi and Houston, Texas. On May 1, Troop H embarked with some relief for Manzanillo, Cuba as part of the occupation forces. Still, their challenges did not end when they left the southern United States. After less than two months, yellow fever struck the troop barracks. Troop H moved four and a half miles to a sugar mill near El Cano where they quartered in "a shed large enough to comfortably hold the troop." The old barracks were closed and all property and personal gear disinfected. William D. Moody, who had just enlisted in March at Louisville, Kentucky, died on the Fourth of July. The rest slowly recovered while the troop continued to perform routine camp duties and drills. On January 11, 1900 Troop H arrived back in Galveston aboard the transport ship Kilpatrick. They had been assigned as one of four troops in the Second Squadron to man outposts along the Rio Grande while the other eight troops remained in Cuba. After traveling by rail to Fort Clark, the troopers transferred fifty-five of their horses to Troops F and G. At that time, H Troop had been officially designated to receive only sorrel horses. Following the transfer, they did pick up thirty-seven additional horses, but only four were sorrel. Captain Hunt wrote that all the moves "have made the troop which was sorrel of mixed color."[198]

Dennis Bell had received his Medal of Honor while serving at Manzanillo. He had just recovered from a serious case of malaria when he received both the medal and a promotion to corporal. Bell evidently wore his medal proudly and frequently. In January 1902, he requested a new ribbon, "the ribbon now in my possession having become unfit for use by fair wear and tear." Even after his enlistment ended, Bell continued to wear his medal. He wore it when he visited Lewis M. Smith, former First Sergeant of Troop M at Alexandria, Virginia early in January 1906. Bell lost the original medal on that visit, probably while riding a trolley. He did request a replacement on March 3. Bell continued to live and work in the Washington, DC area throughout the rest of his life. After witnessing many of the remarkable changes of the twentieth century, Bell died on September 25, 1953 and was buried at Arlington National Cemetery. Not all of the Medal of Honor winners at Tayabacao fared as well as Dennis Bell. Fitz Lee received a disability discharge at Fort Bliss on July 5, 1899 and moved to the Fort Leavenworth area where many former soldiers had settled. Charles I. Taylor of Troop H, later a manager in the Negro baseball leagues, and his wife Cora took care of Lee during his last months. Lee died at the Taylor home blind and in great pain on September 14, 1899.[199]

At the same time Dennis Bell and Troop H served at Manzanillo, the troop's old commander Colonel Louis Carpenter also was stationed in Cuba. Commanding a brigade of the Seventh and Eighth Cavalry, Carpenter took control of Puerto Principe from the Spanish on November 30, 1898. Although the Spanish barracks were thoroughly cleaned for the Americans, some cases of yellow fever still plagued them. Carpenter's cavalry scouted the region throughout the following spring. With help from the Rural Guard, they "suppressed all brigandage" and established a local court system and government structure. On March 18, 1899, General Lopez Recio became governor of the province on Carpenter's recommendation. Carpenter was relieved of command on June 24 and arrived in New York on July 30. On August 12, the colonel had an interview with President William McKinley at the Champlain House near Plattsburg, New York. He received his appointment as Brigadier General on October 18 and retired the following day with 38 years of service. Belatedly, the army had recognized his service at the Beecher Island rescue and the action on Beaver Creek with the Medal of Honor presented April 8, 1898. His citation read, "Was gallant and meritorious throughout the campaigns, especially in the combat of October 15 and in the forced march on September 23, 24 and 25 to the relief of Forsyth's Scouts, who were known to be in danger of annihilation by largely superior forces of Indians."[200]

When Troop H arrived at Fort Clark in January 1900 they stood at full strength with eighty-two enlisted men, eleven of whom were sick and five under arrest. The nearby town of Brackettville and surrounding Kinney County only had a male population of 711, mostly cattle and sheep ranchers. A third of that population proved to be the Black soldiers at Fort Clark. In the more isolated conditions Troop H did not experience the high level of discrimination that had plagued them in recent years. In addition to First Sergeant Shelvin Shropshire and Quartermaster Sergeant Charles Faulkner, the troop complement included five sergeants, six corporals, sixty-one privates, two cooks, two trumpeters, farrier, blacksmith, saddler, and waggoner. The average age for a trooper was twenty-eight with half of them being between eighteen and twenty-five. There were five veterans over forty-five. Most of the members of H Troop continued to enlist from the southern United States although eight hailed from Ohio and the largest number, twelve, were from Kentucky. By comparison, white soldiers stationed at Fort McKenzie, Wyoming at the same time tended to enlist at a younger age and about fifteen percent of them had immigrated from outside the United States.[201]

In November 1900, Captain Robert D. Read took command of Troop H. Read, a native of Tennessee, graduated from West Point in 1877 and had been assigned as captain of Troop K of the Tenth Cavalry in Cuba before his appointment to H Troop. The regimental headquarters and eight troops remained in Cuba under the command of Colonel Samuel M. Whiteside, a Canadian immigrant and Civil War veteran. The Second Squadron did not join the regiment in Cuba. They were needed elsewhere. In the spring of 1901, H Troop boarded the trains again for San Francisco with the rest of the Second Squadron. On April 9, they embarked for the Philippine Islands aboard the transport ship *Logan* to assist in combating the insurrection that had risen among the Filipinos. Before the war between Spain and the United States, Filipino nationalists under Emilio Aguinaldo had been engaged in a battle for independence. With the Americans taking control of the Philippines through the Treaty of Paris, the revolutionaries found that they simply had traded the colonial power of Spain for that of the United States. Many U. S. citizens questioned the move toward imperialism, but most agreed that civil order should be maintained.

The Philippine war for independence against the United States began on February 4, 1899. Through most of 1899, Aguinaldo attempted to wage conventional warfare against well equipped, experienced American troops. By the end of the year, he had converted to guerilla tactics, striking U. S. camps and Filipino villages that cooperated with the Americans. That was the prevailing situation when the Second Squadron landed at Manila on May 13, 1901. To make matters worse, they arrived without their horses which were supposed to be shipped on a separate transport. Troop H received assignment to Mao on the island of Samar. Forced to travel by foot or boat around the island, they spent the next several months locating rebel camps and "destroying insurgent supplies."[202]

One night while Troop H camped on the beach, a U.S. Navy ship saw their fire on shore and thinking it was a rebel camp fired a shot at them. The troopers later recalled that by the second shot "all 100-yard records had been beaten." Still without horses, the soldiers beat their way through many areas of Samar where they had difficulty even walking. As they continued to capture insurgent camps, the troop did seize "about seventy-five small ponies" which they used as pack animals. They also captured fugitives and used them for manual labor to repair boats and bridges. Captain Read reported, "When the troops first arrived in this district May last, the natives would invariably take to flight like frightened rabbits on the approach of the soldiers, but after the severe punishment administered after the first three weeks of operations here, during which great quantities of supplies and shelter was destroyed, one native killed, several wounded and about thirty made prisoner while trying to run away; the natives began to present themselves and be registered, renouncing all connection with the insurrection and proclaiming their submission to American sovereignty and in obedience to the orders of the district commander bringing in their families from the woods and rebuilding their homes along the coast..."[203]

With the rebellion under control on Samar, the squadron moved to the island of Panay in August with Troop H assigned to the outpost at Pototan. They continued to spend much of their time in the field although many of the rebels had surrendered and returned to their homes. In September, they received replacement horses for those that had never been shipped from San Francisco. Unfortunately, the remounts were plagued by disease before they could even be trained as cavalry horses. In December, Troop H lost twenty-two horses, two of them shot on December 31 to prevent further contagion. They lost an additional twenty-eight mounts in January. Most of the horses died of surra, a protozoa infection carried by horseflies. Others succumbed to glanders, a bacterial infection usually found in contaminated feed or water. The cavalrymen gave their horses the best possible care and attempted to improve conditions, but the animals continued to suffer casualties over the next few months.[204]

Even though there was still sporadic fighting, on July 4, 1902, President Roosevelt declared an official end to the Philippine insurrection and began to work toward more self-government and economic development. The Second Squadron, Tenth Cavalry had already sailed from Manila on June 28 bound for San Francisco after a stop in Japan. In San Francisco, they boarded trains for Fort Mackenzie in northern Wyoming where they would be stationed for almost four years. Fort Mackenzie had not been built until 1899 with the stated purpose of protecting settlers from the large Indian population around Sheridan. In reality, it was generally considered to be a political gift to the new state of Wyoming from their congressional delegation rather than a defensive installation. It certainly proved to be a far different environment for the Black troopers than guerrilla fighting in the Philippine jungles. Sheridan was a small ranching community consisting mainly of white settlers from midwestern states including Iowa, Nebraska and Illinois. The Buffalo soldiers constituted almost the entire Black population in that part of the state. Local citizens tended to segregate the soldiers of the Tenth and treat them in the same unjust manner that they treated other minorities in the local area including the Shoshones, Cheyennes, and Arapahoes. A newspaper in the nearby town of Big Horn had stated, "The better class of citizens are determined to see that the color line is drawn at least in regard to who shall occupy buildings in the principal business part of town."[205]

During this period, the headquarters of the Tenth Cavalry moved to Fort Robinson, Nebraska where Colonel Jacob A. Auger, son of Civil War General Christopher C. Auger assumed command of the regiment on June 9, 1902. Captain Read was promoted to major in January 1903 and Captain Charles T. Boyd became commander of Troop H. Boyd had been born near Sperry, Iowa on October 29, 1870. He taught school for a brief time before attending West Point, where he graduated in 1896. After being stationed at Fort Leavenworth with the Seventh Cavalry, Boyd went to the Philippines with the Fourth Cavalry. During the insurrection, he served as a major of volunteers recruiting men who had been discharged in the islands for further service and commanding a battalion of Philippine scouts. Returning to the United States, Boyd received his commission as captain in the regular army and assignment to the Tenth Cavalry. During that time he married Carlotta Klemm and they had two children.[206]

Life on post proved calm after service in Cuba and the Philippines. Only one soldier, Charlie Jones, deserted from Troop H at Fort Mackenzie. In December 1903, Medal of Honor recipient Dennis Bell, still suffering from his bout of malaria received a disability discharge. The troopers spent much of their off-duty time playing sports including baseball and basketball. In a track and field competition sanctioned by the War Department in 1903, Private Seth Andrews and Corporal Lloyd Stafford of Troop H came in second and third respectively in the 220-yard run, while Private Andrew Knight placed second in the broad jump. While sports provided recreation and relief from the daily routine, it also opened new opportunities for Black men in the military and afterward in civilian life. During this period, baseball's first "Colored Championship of the World" was played in 1903 and in 1904 the first Black basketball teams formed in New York and Washington, DC. Charles I. Taylor, after serving in Troop H during the Spanish-American War, founded a Black professional team, the Birmingham Giants in 1904. He became a successful manager and second baseman, helping to found the Negro National League in 1920 and serving as the league's vice-president.[207]

Captain Boyd spent most of this period on detached service as Professor of Military Science and Tactics at the University of Nevada. That left Troop H under the direct command of New Jersey native, Lieutenant James S. Greene. Unlike most of his predecessors who had experience on the frontier or in a war zone, Greene graduated from West Point in the spring of 1904 and took over Troop H in June. In October 1906, the troop moved to Fort D.

A. Russell near Cheyenne. While there the four troops of the Second Squadron prepared to return to the Philippines as part of an army of occupation. They made a rendezvous with the rest of the Tenth Cavalry and embarked from San Francisco on March 5, 1907.[208]

Reaching Manila on April 2, the Tenth Cavalry settled in to nearby Fort William McKinley. With no more rebels to chase, Troop H soon became involved in sports again with Private James J. Prather taking first in the one-mile run, while Private George Coleman won the broadsword competition. In the officers' division, Second Lieutenant Emmett Addis took first in the one mile run and Lieutenant Greene placed second. The troop also experienced several changes in command structure. Veteran Quartermaster Sergeant Charles Faulkner received appointment as regimental Sergeant-Major on September 1, 1907 and retired in October after twenty-eight years of service. Captain Boyd had been assigned Assistant Adjutant General for Fort McKinley, so on June 4, 1908 Captain Eugene P Jervey, Jr., a former West Point instructor in Civil and Military Engineering assumed command of Troop H. Jervey had recently returned to duty after being hospitalized in San Francisco. He continued to suffer poor health while Lieutenant Greene still maintained the daily troop operations.[209]

In October 1908, H Troop marched inland along the Marikina River making camp at San Mateo in Rizal Province. The Marikina River provided the main water supply for the Manila area. Since the cholera epidemic in 1902, American troops had patrolled the river watching for any signs of disease or contamination. Troop H rotated through several month-long patrols during their assignment in the Philippines. On April 26, 1909, Captain Jervey died without ever recovering enough to take active command of the troop. That same month, Colonel Jacob Auger died of apoplexy. The new commander of the Tenth cavalry was Colonel Levi P. Hunt, who had captained Troop H from 1894 to 1900. In May 1909, the Tenth boarded the transport ship *Kilpatrick* to return to the United States. Instead of crossing the Pacific to San Francisco, the transport sailed west through the Suez Canal and across the Mediterranean and Atlantic to New York harbor. The Buffalo soldiers of the Tenth had traveled around the world, but there were still new frontiers ahead. Their next station would be in a part of the country where the regiment had never been before, Fort Ethan Allen in Vermont.[210]

Charles I. Taylor, Troop H
Negro League baseball manager

THE NEW CENTURY 1909-1931

George Osborne, Troop H in Vermont.

On February 12, 1909, the National Association for the Advancement of Colored People organized in New York City. During that same time period, Jack Johnson became the first Black man to earn the title, Heavyweight Boxing Champion of the World. His father had been a teamster with the United States Colored Troops during the Civil War. On April 6, the Arctic expedition under Robert Peary reached the site they believed to be the North Pole. Peary's field assistant, Matthew Henson, a Black man, was in the lead. On February 22, President Roosevelt's Great White Fleet led by sixteen battleships returned to Hampton Roads, Virginia after circumnavigating the world on a stated mission of power and goodwill. The Tenth Cavalry had travelled almost as far since leaving San Francisco in 1907. On July 25, 1909, their transport ship, the *Kilpatrick*, entered New York harbor. After seventy days on the ocean, the cavalrymen arrived on land to find that the city of New York had prepared a ten-mile parade route through the city in their honor.

Hailed as heroes of San Juan Hill, they marched off the *Kilpatrick* on the morning of July 26 to "a noisy welcome and a hearty one ..." The regiment marched down Wall Street and up Broadway to the New York City Hall where they were reviewed by the mayor, George McClellan, Jr., son of the Civil War general. They were accompanied in the parade by "mounted police and two or three bands, a colored vaudeville association in long grey dusters, a double line of twenty carriages or more, carrying the reception committee, made up of prominent Negroes of New York and New Jersey." The Tenth continued up Broadway and Fifth Avenue until they came to the Sixty-Ninth Infantry Armory. A special luncheon had been prepared for them in the vast, open drill hall. As they marched toward the armory from Fifth Avenue, Private George Coleman of Troop H, a cook from Virginia collapsed in the heat and had to be transported to the hospital. The President of the Board of Aldermen, Patrick McGowan began the lunch program by praising the regiment saying "you men have raised your race higher in the esteem of the people of the United States than it ever was before. You have done more to kill the prejudice of some of our citizens than anything else ever could... If the state of Vermont doesn't want you, come back to us by all means." A choir of Black school children sang while the weary troopers enjoyed the feast.[211]

The following day, the Tenth Cavalry arrived by train at Burlington, Vermont and marched into Fort Ethan Allen located near the town of Colchester. Built in 1894 to defend the border with Canada, the post proved to be more modern than any the troopers previously had occupied. The buildings were electrified and a trolley line extended to the post in 1905. Fort Ethan Allen also boasted an indoor riding hall for conducting cavalry drills and athletics during the winter. The first winter in Vermont still proved to be very difficult for soldiers who had previously served in the tropics. George Osborne of Troop H remembered that through the first autumn soldiers were still wearing the summer uniforms that had been issued to them in the Philippines. When 1910 began, Osborne and his fellow troopers "like to froze to death." In February 1910, the Burlington area recorded one of their heaviest months of snowfall with 27.6 inches. Guard tours and company inspections were conducted outdoors even during blizzards. The indoor drills and marksmanship competitions with monetary awards helped the troopers survive an otherwise grueling winter.[212]

The arrival of the Tenth Cavalry at Fort Ethan Allen increased the Black population of Chittenden County exponentially. In the 1900 census only 117 Blacks lived in the county. In 1910, the census listed 898 Black males in Chittenden County, with only 975 adult Black males in all of Vermont. Including family members and camp followers, the total number of new Black residents in the Burlington area was over 1200. Many white citizens initially expressed concern that the sudden influx of Black soldiers would cause unrest. Some even suggested instituting Jim Crow facilities with separate trolley cars or restricting the soldiers to their post. The newspaper editor in Rutland, Vermont while opposing the transfer of Black troops to the state asserted that "the troopers of the 10th will have to conduct themselves twice as well as white soldiers." Those objections were quickly overcome when local citizens began to observe the troopers on a daily basis. Townspeople were invited to watch daily drills and the weekly parade on the post parade ground. The regimental band played public concerts every week under the direction of Alfred Jack Thomas, formerly a member of Troop H. Thomas had completed a course of study with the National Conservatory of Music while in the Philippines and was appointed bandmaster of the Tenth Cavalry in 1909 making him one of the first Black band leaders in the army. His band performed frequently and enthusiastically helping to increase the positive opinion held toward the Black regiment by the citizens of Vermont. By August 14, even the Rutland editor wrote in his column that "the state of Vermont made a good exchange when the hoodlum white [soldiers] . . . were replaced by negro cavalrymen."[213]

During that period, the rate of desertion in the Tenth Cavalry remained low at 1.5 percent compared to white regiments that averaged 3.7 percent. In the years 1910 and 1911, no soldiers deserted from Troop H. The majority of troopers in the regiment in 1910 were between the ages of twenty and thirty with only a few being under twenty or over fifty. Most of them still hailed from the southern states. A significant number of the men had been recruited from the states of Georgia, Tennessee, Virginia, and Kentucky. Others came from areas where the regiment had been stationed, like Texas, with only a few soldiers listing their homes in states such as Kansas, Illinois or California. The average time in active service was a little over five years which meant most of the soldiers had re-enlisted. Thirty-three men boasted more than twenty years of service with the longest tenure belonging to William Thacker of Lawrenceburg, Kentucky. Thacker retired as First Sergeant of Troop H on December 11, 1912, at the age of fifty-one.[214]

Tenth Cavalry on March, 1912. From the collection of the Smithsonian National Museum of African American History and Culture.

In the spring of 1910, the troopers began conducting practice marches in the local countryside to break the monotony of drills on post. In July they marched about 180 miles to Pine Camp, New York, a training ground just east of Lake Ontario. They performed maneuvers including mounted drills with pistol and saber until August 30 when they returned to Vermont by rail. In addition to returning to Pine Camp several times, the Tenth also conducted summer maneuvers near Berkshire, Connecticut in 1912. When they were at the post, the officers encouraged the soldiers to continue their education with regular classes in the evening and a post library. They even established a literary society that debated topics such as "Resolved, That Women are of Greater Value than money." After Thanksgiving dinner in 1911, the troopers staged a Wild West show in the riding hall. The performance included bucking horses, bareback riders and an attack by Indians on a stagecoach and settler's cabin, ending with the Indians being driven out of the hall by soldiers of the Tenth Cavalry.[215]

During the first winter, a post basketball league organized with one team from each company and support organization. After that the soldiers played basketball almost every evening. In February 1911, the regiment sent a special team to compete against the New York All Stars, a civilian all Black team. The Buffalo soldiers lost to the All Stars, 30 – 14 after a strongly contested and well attended game. On evenings when they did not play basketball, the soldiers watched silent films of fights between boxers including Joe Gans and Jack Johnson. In the spring, baseball became the most popular sport on post. George Osborne of Troop H recalled spending all of his off-duty time in warmer weather playing baseball for the company team. The troopers frequently played civilian teams including Saint Michael's College and the University of Vermont. In 1913, they defeated the Mohawk Giants, a Black professional team, in Schenectady, New York. For a time, they played local teams on Sunday afternoons until a complaint from one of the ministers to the War Department caused the Army to ban Sunday games. Colonel Thaddeus Jones replied that no one was forced to attend the games and there was far more vice outside the post than inside.[216]

Probably the worst display of violence on and off post occurred on October 10, 1911. Matthew Carlyle from Georgia was on his second enlistment with Troop A. On October 10, he failed to report for duty when assigned to a special painting detail. The trooper in charge of the detail, Andrew C. Fox, reported this and Carlyle was arrested when he returned to post from Winooski. He was confined to his barracks where he loudly denounced Fox, claiming that Fox had caused him to be arrested several times before and that this time would probably lead to his discharge. A little after noon, Carlyle left the barracks against orders carrying a carbine and an ammunition belt. He approached the painting detail, aimed his rifle at Fox and shot him in the back of the head, blowing off part of his skull. With members of the detail converging on him, he ran down the main road leading off post, stopping periodically to shoot at his pursuers. He continued about a quarter of a mile further to Bluefort's, a local restaurant frequented by the soldiers. Carlyle entered the restaurant and ran upstairs where Clara Washington, a young woman who may have rejected him, lay sick in bed. He shot her three times, once in the head. People in the dining room downstairs began to scatter. As Carlyle came down the stairs, he saw another young woman, Beatrice Stewart, running down the hall toward the outside door and shot her in the back.[217]

At about that time, the post guard under the command of Captain Edwin M. Suplee caught up with him. Suplee had recently assumed command of Troop H after serving several years as the regimental paymaster. While the Captain had his men surround the restaurant, Carlyle retreated to an upstairs window where he could fire down on them. They continued to exchange shots for about ten minutes until the soldiers got a good shot at Carlyle by creating a distraction with part of the guard. With a bullet through his thigh, Matthew Carlyle finally surrendered. The incident caused some initial outcry against the Black soldiers. On October 11, *The Norwalk Hour* in nearby Connecticut claimed that "the protest against sending the negroes here was strong" and that "Today residents are demanding that their Washington representatives have the negroes replaced with whites." If there were any such demands, they soon died out, probably because all of the victims were Black and as atrocious as the incident may have been, it did not directly affect the white communities. Carlyle stood trial in a federal court in February 1912. Although he was convicted of multiple murders, he received a life sentence, largely because his attorney had been able to convince the court that insanity ran in his family.[218]

Life at Fort Ethan Allen returned to the routine, but in 1913 the summer maneuvers proved to be more extensive than usual. The regiment made a march of almost 720 miles to establish a Cavalry Camp of Instruction about five miles outside of Winchester, Virginia, in the Shenandoah Valley. The Eleventh and Fifteenth Cavalry regiments joined them there to drill using the latest cavalry tactics and performing joint maneuvers. At the end of September, all three regiments moved to Fort Myer, Virginia, near Washington, D.C. On October 7, they drilled at Potomac Park and two days later they demonstrated their techniques for President Woodrow Wilson, General Leonard Wood and other dignitaries from the War Department and Congress. Local Black citizens held a reception for the Tenth at which General Wood spoke reminding them that their regiment represented "the colored race, and the eyes of all are upon it." After such an extended period in the saddle, the troopers left Washington by train, arriving back in Vermont on October 12.[219]

Members of Troop H at Fort Ethan Allen, Vermont.

Soon after that, the Tenth Cavalry received orders transferring the entire regiment to Fort Huachuca in southern Arizona. They left Fort Ethan Allen on December 8, 1913, traveling by rail to New Jersey. The *Burlington Free Press* reported their departure recalling that the Black troopers had been "always courteous and gentlemanly." Arriving in Weehawken, they once again boarded the now aging transport ship *Kilpatrick* bound for Galveston, Texas. Lieutenant John B. Brooks remembered that "The men had to unload the freight trains and do practically all the stevedoring to put the cargo aboard the ship." He described the voyage to Galveston as "very cold, windy and rough." The men then had to unload the ship onto cars of the El Paso and Southwestern Railroad. The regiment finally arrived at Fort Huachuca on December 19.[220]

In Arizona, Troop H returned to the duty of patrolling the Mexican border they had left almost twenty-five years before. Many of the recent recruits had never been under fire or spent extended periods in the field. Almost as soon as they arrived, the Troops H, G and M departed for Nogales, southwest of Fort Huachuca along the Mexican border. From there they established camps and patrolled the border area all the way from Nogales to

Yuma about 300 miles to the west. They also patrolled east at least as far as Naco, Arizona. Their stated mission was to ensure that U.S. citizens observed neutrality laws and to enforce the arms embargo, that was periodically lifted, depending on who held the political power in Mexico. Because of the ambiguous nature of the embargo on firearms, U. S. arms dealers continued to smuggle weapons across the border, even with the presence of cavalry picket lines. The thin spread line of soldiers was welcomed by local ranchers who had repeatedly lost livestock to Mexican raiders.[221]

Across the border in Mexico, the governor of Sonora, Jose Maria Maytorena attacked Mexican government forces in Naco in October 1914, resulting in a 119-day siege of the city. Troop H was part of a detail sent to reinforce the U. S. side of the border and prevent any Americans from participating in the battle. The troopers spent much of their time on duty in trenches and rifle pits guarding the border, but not allowed to return fire from the Mexican combatants on the other side. George Osborne recalled, "My work in Mexico was guard duty mostly, straight duty they called it. And otherwise there was fighting, but we didn't get no credit for it." On December 4, Private John W. Miller, who had just joined Troop H from the recruiting depot in September received a flesh wound four inches below the groin. The wound proved serious enough that he received a disability discharge after only a few months of service. Two days later, Sergeant Charles Smart of Troop H was shot in the toes of his left foot. When the fighting ended in February 1915, President Woodrow Wilson sent a message to the Tenth complimenting them on their discipline and their enforcement of the neutrality laws.[222]

Returning from Naco, Troop H once again settled into the routine of life on post. While the border camps had been remote and the soldiers often lived in tents, Fort Huachuca boasted electric lights and phone service and provided a library, bowling alley and swimming pool. Daily life was carefully regimented. Troop orders for 1915 directed that the non-commissioned officer in charge of quarters could not leave the barracks for any purpose and "coming off duty will get the mail and if not at drill will take the Sick Report to the hospital." Care of the troop horses included walking them for the first mile out and last mile in. Horses were not to be ridden at a gallop except in drill and no horses could be turned loose in corral.

Troop H did return to the border later in 1915, where in November they participated in a skirmish at Nogales against the forces of Pancho Villa across the border.[223]

While at Fort Huachuca, Troop H took part in the 1916 campaign against Mexican rebel Francisco "Pancho" Villa. During the period of continuing political unrest in Mexico, Villa had rebelled against Mexican President Venustiano Carranza and gained control of most of northern Mexico. On January 10, 1916, Villa stopped a train with personnel from the Cusi Mining Company at Santa Ysabel between El Paso and Chihuahua and executed sixteen miners. Two months later on March 9, he led over 500 men across the border to raid Columbus, New Mexico killing five and wounding many others. General John J. Pershing, formerly of the Tenth Cavalry, was appointed to lead a punitive expedition across the border to capture Villa and end his threat to United States citizens.

Two columns entered Mexico on March 15, one consisting of the Eleventh and Thirteenth Cavalry and the other the Seventh and Tenth Cavalry. The Tenth had marched from Fort Huachuca to Culbertson's Ranch in the southwest corner of New Mexico to assemble their troops for the campaign. First Lieutenant Orlando C. Troxel of Troop H recalled that "only now and then did we have hay, watering facilities were always poor, the supply insufficient and frequently none except at our nightly camps, and the country was sand and devoid of grazing. We thus marched 160 miles before we entered Mexico. We lost several horses from sand colic and all horses had begun to feel the effects of the march." Conditions after entering Mexico did not improve. Although President Carranza had not tried to prevent Pershing from crossing the border, he still did not support the incursion and refused to allow the U. S. troops the use of the Mexican Northwestern Railroad to carry much needed supplies. Because of a shortage of supplies locally, even when Pershing willingly paid the Mexicans, the army tried using trucks to convoy supplies from the border. That proved to be the first successful use of a truck convoy in military operations.[224]

On April 1, F and H Troops of the Tenth Cavalry discovered 150 Mexican soldiers at a collection of ranch buildings called Agua Caliente. Commanded by a Buffalo soldier, Major Charles Young, they advanced in line under the cover of machine gun fire from a battery overlooking the compound. The Mexican troops quickly scattered before the charging horses and troopers with their pistols drawn, leaving three dead. The engagement at Agua Caliente was the first use of machine guns in combat by the United States. Lieutenant Troxel claimed, "We never saw these Villistas as opponents again." He continued, "None of our men were hit and the horses were the only part of our command that had not enjoyed the skirmish. One horse was wounded, one of mine dropped exhausted, one died that night, we killed one the next morning, and one could just get along by being led. I do not know the loss of the animals in other troops." Back in Vermont, when the citizens of Winooski heard about the engagement in Mexico, they wrote to the commander of the Tenth offering "the sympathy, gratitude and appreciation of all of the citizens of this community because of these achievements of men who were our neighbors and friends and who met this supreme test and sacrifice in a manner to thrill and inspire every true American."[225]

While the objective of the United States troops was to capture Pancho Villa, the Carranza government openly objected to their presence in Mexico and Carranza troops clashed with Pershing's expedition at Parral in April and at Carrizal on June 21 where Captain Charles Boyd, in charge of a detachment of about one hundred soldiers of the Tenth Cavalry, was killed. After that Pershing's expedition went into camp at Colonia Dublan in northern Chihuahua in an attempt to prevent a war with the Carranza government while still pressing them to capture Villa. By the end of the year, it was obvious that the soldiers would see no further action so a Christmas feast and celebration was planned. The feast was spoiled by a cold and violent windstorm that coated the cattle being barbequed with a layer of dust and forced the soldiers to find any available shelter. Finally in February 1917, with war against Germany imminent, the troops left Mexico and Troop H returned to duty patrolling the border with each detachment covering about twenty miles.[226]

When not on patrol, the men dug trenches and began to practice trench warfare, grenade throwing and use of gas masks in preparation for war in Europe. The American Expeditionary Force organized in July, but this time the Tenth Cavalry did not join General Pershing. The War Department's final decision was not to send the Black regulars to Europe. That decision may have been affected by an incident in Houston on August 23, 1917, when close to one hundred men of the Twenty-Fourth Infantry marched into the Black San Felipe district of Houston to avenge the beating of two of their battalion by policemen earlier that day. The Third Battalion of the regiment had been stationed in Houston to guard construction of the training facility at Camp Logan. The soldiers frequently visited the Black district in off duty hours. Close *by was the Houston red light district and police trying to control traffi*c apparently took out their aggression on the Black soldiers. The infantrymen marching to avenge their comrades apparently shot at anyone who surprised them, killing fifteen, four of them policemen, and seriously wounding twelve. Four of the soldiers died. Those remaining on post formed a picket line to protect against expected retaliation by a white mob. No company roll was taken so there was no certainty of who was involved in the march. One hundred eighteen men were charged with mutiny and riot and nineteen were executed by hanging.[227]

Because most of the local ranchers and businessmen in Arizona appreciated the Tenth Cavalry's defense of the border, racial problems did not affect Fort Huachuca during this period. Although the regiment did not leave for Europe, up to a third of its members were assigned to volunteer all Black units to provide them with experienced leadership and training. Sixty-four of the Tenth's non-commissioned officers received commissions during the war. From Troop H, First Sergeant Clifford A. Sandridge, a twenty-year veteran from Ohio was appointed Captain and assigned to the Muese-Argonne sector. After the war, he continued to serve as a commissioned officer until his retirement in 1932. Other members of Troop H commissioned from the Des Moines training camp on October 15, 1917 were Corporal General Lee Grant from Mississippi as a second lieutenant and Sergeant George E. Edwards as a first lieutenant. When those new officers left for France, many of the white officers in the Tenth Cavalry also transferred to units going overseas. Troop H came under the command of Captain Pinkney Armstrong, a veteran sergeant from the Fifth Cavalry who had been granted a commission with the regulars for the duration of the war.[228]

For the remainder of 1917, the Tenth Cavalry continued their patrols of the Arizona border. Troop H spent many months in the field in rotations with other troops including Troop E using the camp near Nogales. Lieutenant Harold Wharfield later recalled "spending our spare time hunting while patrolling for smugglers and bandits, as well as Yaqui Indians, from across the line. It was hard to imagine better duty." During the fall, ranchers near Nogales reported missing livestock and armed Yaqui Indians crossing the border. They requested that the Army increase its patrols in the area. The Yaquis had entered the United States to find jobs picking fruit and cotton in the Salt River Valley of southern Arizona. Many of them would then use their wages to buy arms and ammunition and smuggle them back into Sonora. The Yaquis had been engaged in a decades long war with the Mexican government over independent control of their homeland in Sonora. During the late nineteenth and early twentieth century thousands of the Indians were sold into slavery on the sugar cane and tobacco plantations of southern Mexico. While the Black soldiers of the Tenth Cavalry might have sympathized with the plight of the Yaquis, one of their primary directives continued to be the prevention of arms smuggling across the border.[229]

At the beginning of 1918, Troop E established a camp next to an old homestead in Bear Valley about thirty miles west of Nogales. Yaquis had been sighted crossing through this uninhabited area in an attempt to avoid the patrols. On January 9, about thirty Yaquis fired on a detail from Troop E under the command of Lieutenant William Scott believing them to be Mexican troops that had crossed the border. Scott's patrol pursued and captured ten of the Yaquis, mortally wounding their leader. The Mexican government insisted that the captured Indians be returned to them for trial. Instead the United States tried them for illegal exportation of firearms and sentenced them to thirty days in jail. One eleven-year-old boy was released. The day after the Yaqui fight, Troop H arrived to relieve E Troop on border patrol at Bear Valley. The outfit that had once boasted some of the most experienced Indian fighters on the frontier had just missed the last Indian battle on American soil.[230]

Though local ranchers and other citizens frequently had praised the Black troopers, with the end of World War I and a

146

steady decrease in the threat from Mexico, racial tension became more pronounced. Many of the nearby settlements did not welcome members of the Black Tenth. One racial incident did occur in Bisbee, about thirty miles east of the post on July 3, 1919. The Tenth Cavalry was in town to participate in a Fourth of July parade the next day and many of the troopers decided to enjoy an evening out. The trouble began with an altercation between several soldiers and a white provost marshal in Brewery Gulch. The sheriff decided to disarm all of the troopers in town and a fight broke out resulting in three soldiers, a deputy sheriff, and a young woman being wounded. Fourteen soldiers were arrested. None of the injuries proved serious and the regiment marched in the parade the following day. While the incident demonstrated the potential for racial problems, there was no demand for the removal of Black soldiers. Troopers continued to visit Bisbee on their off-duty time because of the small Black community developing there. Tombstone proved more stringent in their enforcement of Jim Crow laws and while Tucson was a larger city with more possibilities, it was nearly eighty miles away over a road that "was dirt and narrow, filled with chuck holes, puddles, and all manner of obstruction." The White City, a small settlement with stuccoed white walls just outside the post gates consisted mainly of saloons and brothels. Despite warnings from the officers, that settlement became the most popular destination for off duty Black soldiers. Lieutenant John Brooks recalled, "It was my understanding that the girls there were largely white girls when the white soldiers were here, but when the 10th Cavalry came, why the white girls left and the colored girls came in very promptly. I think they were probably on the next train behind us." [231]

In 1920, there were 1178 Black males in Cochise County, most of them at Fort Huachuca. The entire Black male population of Arizona totaled only 5859. Compared to a total male population in Cochise County of 12,744, the Black troopers represented a true minority and may have frequently felt isolated from society. The majority of troopers in the Tenth during this period were between the ages of twenty and thirty. There proved to be more soldiers between eighteen and twenty than at any time previously since a number of young men had been recruited to serve during World War I and then had chosen to fill open slots in regular regiments instead of mustering out at the end of the war. Some like Atwood Mitchell and Garfield Dyas had served with the 312th Service Battalion in France. There were also still some old soldiers like forty-eight-year-old George A. Lee who had served with Troop H as a bugler. Apparently, the oldest serving trooper in 1920 was Howard W. Smallwood who at age fifty-seven served Troop L as a cook. The majority of the soldiers still hailed from southern states like Georgia, Alabama and Kentucky although the Tenth also had recruits from a diversity of states including Connecticut, New Jersey, Illinois, Colorado and Utah. At least one came from Mexico and another from the British West Indies.[232]

The Tenth Cavalry managed to survive cuts imposed by the National Defense Act of 1920. Many whites still did not favor integration, but they did think Blacks should be required to serve in the army. On January 22, 1921 the First Cavalry Division organized as part of the new defense act with headquarters at Fort Bliss, Texas. The Tenth was assigned to the division along with the First, Seventh and Eighth Cavalry Regiments and remained at Fort Huachuca throughout the 1920s. Life on post settled into a routine of drills, guard duty and training exercises. For entertainment, the post hosted movies at least once a week. The soldiers would supplement this by performing their own shows. Lieutenant Brooks later remembered that "They liked to sing and they liked to have people come and applaud." Holidays continued to celebrated with feasts prepared by each troop, especially at Christmas. Cornelius Smith, Jr., son of the post commander recalled, "Officers, frequently accompanied by family members, would be guests of troop messes where toasts were drunk and pleasantries exchanged. Commanders would judge the artistic and culinary artistry of troop Christmas dinners, a competition which kept chefs and pastry cooks in a high state of excitement during the holidays. Soldiers called upon troop commanders for "remarks," wherein the witticisms expressed resulted in all manner of railery, applause, hooting and hollering in a spirit of genuine good fellowship."[233]

Athletics continued to be a major source of recreation with baseball as the favorite sport. In 1920, a combined track and field team from the Tenth Cavalry and Twenty-fifth Infantry won the Southern Department track meet. The Tenth and Twenty-fifth also frequently competed in maneuvers and training exercises with the infantry on offense and the cavalry practicing delaying actions. Many of the competitions served the dual purpose of recreation and preparing for combat. Horse shows were staged several times a month for competition and training. The soldiers were required to qualify in the use of pistols, rifles, automatic rifles and sabers. Rifle and pistol competitions became very popular and in 1921, the Tenth Cavalry won the Eighth Corps firearms competition. Rivalry between the Tenth and the Twenty-fifth became even more pronounced when the Twenty-fifth Infantry was assigned to Fort Huachuca in 1928.[234]

In 1926, as aircraft continued to improve and their uses in warfare expanded, the army decided to increase manpower in the Army Air Corps. Because there was no room for overall expansion in the peacetime army, the decision was made to reduce the size of existing units to provide more recruiting slots for the Air Corps. The four Black regiments were to be cut in 1931, the fifth year of reductions. Since Black men were not eligible to join the Army Air Corps, this meant fewer Blacks would be recruited into the army. By the time this reduction in force took place the United States was in the midst of a depression. In every other economic downturn since the Civil War, the army had offered opportunity and a chance for Black men to improve their circumstances. During the Great Depression, that opportunity did not exist. The National Association for the Advancement of Colored People and Black newspapers across the nation protested the lack of chances to enlist for the Black communities. American Legion posts and other civic groups wrote their congressmen expressing their concern for the future of Blacks in the military. Doctor Robert R. Moton, President of Tuskegee Institute, wrote to President Herbert Hoover stating that "the War Department makes no reference whatever to the Injustice of reducing the total number of Negro troops in order to make provisions for a department of the service to which negroes are not admitted ... the War Department completely ignores the representation that negro soldiers are being assigned exclusively to service units; that they are being distributed throughout the country solely to do menial work for white soldiers... without chance for training or promotion and be excluded from other branches of the services." While Doctor Moton's efforts did not succeed in improving the War Department's policy toward the Black soldiers, ironically, when the first Black pilots were recruited during World War II, they trained on Moton Field at Tuskegee Institute.[235]

In October 1931, the Tenth Cavalry was dismounted and reassigned as service detachments at various army schools. Captain Vance Batchelor remembered, "One day the inevitable happened, the War Department decided that the horse cavalry was outmoded. One of the first regiments to be disbanded was the colored tenth. We loaded the horses on the freight cars and sent them away. Next day the Pullman sleepers backed quietly into the rail yard. The colored troops boarded the train with tears in their eyes and started their trip north ..." According to John B. Brooks "It lost its identity as a real soldier outfit when they became nothing but a bunch of dog robbers. And it hurt them too; they didn't like it at all. The old soldiers, as soon as they could retire, instead of staying; they just retired." Troop H traveled as part of the Second Squadron to West Point, New York where their primary duties included the training and maintenance of the horses and acting as orderlies and servants for the instructors and staff. The First Squadron of the Tenth received assignment to Fort Leavenworth and the Third Squadron went to Fort Riley, both in Kansas. Throughout the 1930s, they continued to function as service units with little opportunity for combat training or advancement.[236]

Finally on January 30, 1941, the Tenth Cavalry reassembled as a combat unit at Fort Leavenworth. In 1942, they became part of the Second Cavalry Division along with the Ninth Cavalry and were sent to the Cavalry Training Center at Camp Lockett, California for intensive combat training. Both regiments sailed for the Mediterranean and North Africa in the spring of 1944. Upon arrival, the regiments disbanded and were reassigned as combat support troops in North Africa. Claiming that cavalry was obsolete, the army shuffled them into menial duties where the soldiers had little opportunity to demonstrate their combat readiness. Many of the soldiers were assigned to duty repairing roads or driving trucks and other military vehicles. For some reason, mechanizing the Buffalo Soldier regiments and retraining them was not even considered at the time. Some members of the Tenth Cavalry requested detached service and joined the 92nd Infantry Division in Italy as replacements. As part of the Fifth Army, the 92nd became the only Black infantry division to see combat in Europe. By the end of World War II, the Tenth Cavalry regiment had ceased to exist. In 1950, the Tenth did receive a new designation briefly as the 510th Tank Battalion. Then finally on June 25, 1958, the Tenth Cavalry once again became an official unit of the United States Army. Troop H again stood ready to serve their country, now as part of a totally desegregated force.[237]

Captain Pinkney "Pink" Armstrong

THE LAST CAMPFIRES OF THE BUFFALO SOLDIERS

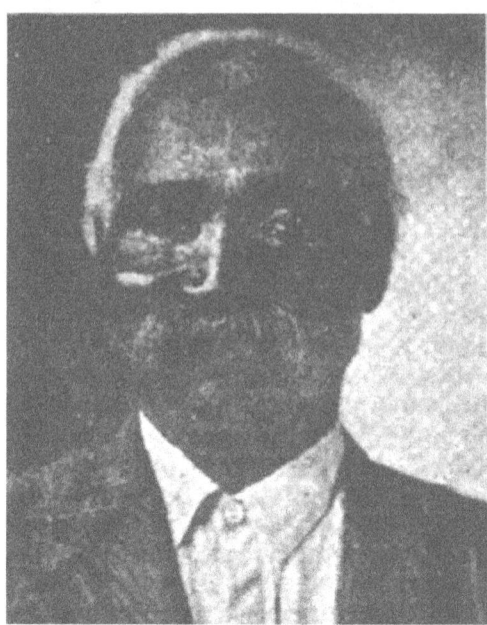

Reuben Waller
Passed away at 105 years of age.

Gradually the frontier passed away as farmers, ranchers and developers moved into areas previously considered desert or wilderness. Cities and towns continued to grow where once only isolated army outposts stood. Railroads crossed the continent. The buffalo soldiers of the Tenth Cavalry had helped to make this possible, but as the twentieth century advanced, their period of useful service was coming to an end. The automobile quickly replaced the need for horse drawn transportation. With ever improving highways, the army began to transport troops and supplies by truck instead of wagon. By World War II, even the cavalry had become mechanized. The old soldiers of Troop H, now long retired, would probably not have recognized the army that embarked for Europe and the Pacific in 1942.

Many of the old soldiers who had served in Kansas, Oklahoma, Texas, and Arizona lived well into the twentieth century. On a late summer day in 1925, "a big automobile rolled up" to the home of Reuben Waller in El Dorado, Kansas. Out stepped J. J. Peate responding to a letter written by Waller to the Indian Wars veterans' periodical, *Winners of the West*, asking if there were any survivors of the rescue at Beecher's Island. Peate had homesteaded in Beverly, Kansas, worked in the creamery business and managed the Martin Lumber Yard. He had achieved some measure of prosperity, purchasing a share in the lumber yard and serving as a shareholder and director of the Beverly State Bank. The veteran scout and the aging trooper had a long visit remembering their days on the frontier. In September 1928, they both attended a sixtieth anniversary commemoration of the battle at Beecher's Island. Waller delivered the opening address to about two thousand people and Peate helped raise funds for a battle monument to be dedicated the following year. Waller continued to write letters to *Winners of the West* attempting to locate members of his old company. In June 1929, he visited Thomas Murphy of Corbin, Kansas, the only survivor of the Forsyth Scouts he could find.[238]

Reuben Waller raised six children and outlived four wives. By 1880, he had settled on a 160-acre farm near Towanda, Kansas with his first wife, Susan Saulters. Their oldest child, Jess, was born September 5, 1888. Jess was followed by Elmer, Ray, Charles, Vera, and Maisy. Waller served for thirty years as superintendent of the Sunday school at the African American Methodist church in El Dorado. By the 1920s, he received a fifty-dollar pension, although he received nothing for his Civil War service having been an "unofficial" Confederate. Concerning the pension system, he wrote, "[W]e were 'regular soldiers' and had to make the west safe for the soldiers of the Civil War to get homesteads in, and $72 per month pensions, while we poor 'regulars' get nothing." Reuben Waller died on August 20, 1945 at the age of 105.[239]

Waller had been wrong, however, to believe that none of the troopers from the Beecher's Island rescue were still alive in 1929. At least one, Trumpeter Silas Jones still lived. Jones had married and settled at Havre, Montana near Fort Assinniboine, the post from which he retired. With his wife, Nellie, he lived there peacefully until his death on July 12, 1936. Sergeant Henry Allen also retired from Fort Assiniboine on October 31, 1897 and remained in Havre applying for a pension in 1913. He had married while stationed in Texas and with his wife Martha had two children, a daughter Lizzie born at Fort Davis and a son Charles born at Fort Buford, North

Dakota. Their companion on many scouting patrols, Pollard Cole, chose not to remain in the northwest and moved to El Paso, Texas with his wife, the former Estephana Gonzales. He died less than three years later on May 20, 1900, while away from home at Georgetown, Kentucky. Estephana spent the next twelve years trying to secure a government pension to benefit their son Joseph Pollard. In 1912, a Mrs. Green of Rancho Verde, California, where Mrs. Cole had been working wrote the Pension Bureau that "in the interim of collecting this red tape, poor Estephana died." Another thirty-year veteran who retired in 1897 at Fort Assiniboine was Corporal Philip Jones. Although originally from Virginia, he also married and settled in Havre. Jones died in 1912 and his wife Mollie began the long process of obtaining a widow's pension.[240]

Soldiers with thirty years of service did receive retirement pay based on the final rank they held. Their widows did not automatically continue to receive those benefits and frequently would only be allowed a pension while they remained unmarried. Troopers without thirty years could receive a disability pension if they could document that they had been injured or had contracted a lasting illness in the line of duty. The majority of veterans from Troop H seemed to have had their requests for disability pension denied when they first submitted their documentation. Many did resubmit their applications and with additional information that allowed them to receive some relief. Aaron Archer's request for pension was denied with the notation "so much thereof as is based on rheumatism and neuralgia, on the ground that said disabilities were contracted since the date of your discharge, as shown by your statement and on the further ground that a pensionable degree of disability from said causes had not been shown since date of filing claim." Alexander Brown of Louisville, Kentucky also received a denial of pension with the note that "you were not honorably discharged from your service in Troop H, 10th U. S. Cav., as shown by a report from the records of the War Department." Brown had stolen weapons and deserted in 1878, yet still felt he deserved a pension. Others proved far more deserving. William Allen, a former shoemaker who had become a minister during his tour of duty with Troop H, served as pastor of Union Baptist Church in Lebanon, Ohio and Zion Baptist in Paris, Kentucky. He married Fannie Clouds on April 21, 1902 at Lexington, Kentucky. Allen applied for a pension because of trouble with his liver from the old wound that had caused his disability discharge and for deafness. He did receive a minimal amount for deafness, but since his gunshot wound not only occurred off duty but outside the post of Fort Davis, the Pension Bureau refused him any further allotment. Allen's condition continued to deteriorate until blood poisoning from his liver affected his mind. He was committed to the Eastern Kentucky asylum for the insane in November 1909, where he died March 10, 1910.[241]

Many of the buffalo soldiers suffered physical problems after years of drills and patrols in every condition imaginable. Washington Hardaway received a disability discharge from Fort Davis in 1880 for a hernia. He became a sailor and later settled in Cincinnati, Ohio. There he continued to have health problems, eventually developing rheumatism and heart disease. Charles Black continued to have problems from his 1870 gunshot to the right eye. When his left eye developed a cataract he could "barely distinguish light" and had to be led around by an attendant. The Bureau of Pensions did reconsider its refusal in Black's case and recommended that his injury be considered in the line of duty and his claim allowed since at the time of the injury "he was in his proper place and doing nothing that he should not do."[242]

Frequently the Bureau of Pensions would question whether injuries actually resulted from military service. Colonel Carpenter had to certify that Sergeant Robert White was disabled in the line of duty. He wrote that White was "a very deserving man" who "contracted piles with which he suffered a great deal. Toward the end of the term of service he was kept on light duty in consequence of his disability." Some of the disputes between veterans and the Bureau lasted for many years. Colonel Miller, working as a teamster and porter in Kansas City, Missouri, applied for a pension because of eye trouble he believed was the result of long scouting patrols in the Arizona sun. The pension bureau wrote the examining surgeon, "Claimant has history of gonorrhea. Is there any evidence of effects of vicious habits?" The surgeon had to certify that Miller's injuries had no connection with gonorrhea before a pension was granted. Asa Weaver, brave leader of the patrol that located Victorio in August 1880, returned to Indiana in 1881 and married Cassie Watkins. A daughter, Myra, was born to them on May 9, 1886. Weaver continued to suffer from problems with his eyesight. On April 14, 1910, Weaver asked if he could meet a pension board of examiners in Rushville instead of Indianapolis stating "I am a poor man and am compelled to labor all I can for support of myself and family." The once strong and capable soldier died impoverished on May 3, 1912 in Carthage, Indiana with his pension claim still pending. His widow Cassie continued to pursue the claim.[243]

Former First Sergeant George Goldsby applied for a pension in 1912. At the time, he lived on a farm near Jennings, Oklahoma posing as a white man by the name of William Scott. His second wife, Effie Henshaw, was white and while most people thought he might have some Black blood in him, they did not question his story. Special examiner for the Bureau of Pensions, William L. Bowie, remarked "I could see that acquaintances did not know what to really believe about it." On further examination, Bowie discovered that Goldsby had deserted from Troop D at Fort Concho in 1878. His statement to the Bureau of Pensions read, "I was accused of being implicated in a fight or riot down there, and I saw that the race prejudice was so strong, that although I was not guilty of the charge, I deemed it advisable to get away from there." Actually, a group of cowboys and hunters had surrounded a sergeant from Goldby's company at Morris' saloon in San Angelo, Texas and cut the chevrons and pant stripes from his uniform. The sergeant returned to post and brought other back troopers armed with carbines into town. During the ensuing fight, one hunter was killed and two wounded while one trooper died and one was wounded. Texas Rangers investigating the incident held Goldsby responsible because he as First Sergeant did not stop the bloodshed. Colonel Grierson questioned the rangers' authority on post and while they debated, Goldsby escaped. Nine soldiers were indicted for murder, but only one received the death sentence and he was acquitted on appeal. None of the hunters were even indicted.[244]

Goldsby quietly crossed into Indian Territory. He eventually married again and farmed in Oklahoma and Kansas. With his first wife, Ellen, he had a son named Crawford Goldsby. Crawford in later years became known as "Cherokee Bill." After a brief career as "one of the most noted outlaws and desperados in the history of the Indian Territory," he was hanged at Fort Smith,

Arkansas, on March 17, 1896. Goldsby's connection to "Cherokee Bill" gave him another reason for hiding his identity. As the son of a mulatto woman and her white master, he had many times successfully claimed to be white, so he became the white farmer, William Scott. Only his need for the funds a military pension could provide caused his identity to be revealed.[245]

Sergeant Michael Finnegan also found himself in need of aid in the early 1900's. He had received a disability discharge in May 1895 for chronic rheumatism. He did receive a small pension for his disability. Finnegan went to sea, working as an Engineer's Steward on ships crossing the Atlantic. In 1914, when war broke out in Europe, he found himself stranded in England without a job. "Ships all laid up except those charted by the gov.," he wrote. His baggage, which had been checked in Cardiff, did not arrive at Liverpool. Not only did he have no clothes or personal belongings, but the certificate he presented to draw his pension was missing. The American Relief Committee of London provided Finnegan passage to New York. From there he went to Philadelphia where he finally made contact with the pension bureau asking that he be sent a pay voucher without showing his certificate "as I am on the verge of starvation" and "am now a pauper."[246]

Other old sergeants of Troop H also suffered from illness and poverty in their later years. First Sergeant Jacob Young settled near Fort Gibson with his wife Polly. He experienced rheumatism and lumbago to such an extent that he could not maintain a job. The pension examiner wrote, "He has been in such indigent circumstances that he was unable to procure the services of a qualified physician to treat the diseases." Young did receive a pension and remained at Fort Gibson until his death on September 22, 1916. First Sergeant John F. Casey, also the son of a slave mother and white father, had been married three times and claimed to have been married a fourth time to the infamous Belle Starr. He had two children, Mary and Frank, with his first wife. For a while he worked at his own barber shop in Saint Joseph, Missouri. Casey, who had once been a sharpshooter, began to suffer from deteriorating teeth and eyesight. With none of his wives left to help care for him, Casey spent the last years of his life at the Military Soldiers' Home in Leavenworth, Kansas. First Sergeant George Garnett received a disability discharge for chronic rheumatism and went to live in Peoria, Illinois. By 1906, he too had entered a Soldiers' Home in Des Moines, Iowa.[247]

Former commanders of Troop H fared far better. Levi Hunt became commander of the Tenth Cavalry on May 11, 1909. He retired as a colonel on August 7 of that year and settled in Saint Louis, Missouri. His son, Lieutenant Claude Hunt, reported his death from paralysis on December 19, 1913. After serving in Cuba and the Philippines, Thaddeus Jones retired as a Colonel of Volunteers. He settled in Long Beach, California where he died March 27, 1939 at the age of ninety. Charles Cooper retired on August 16, 1903 as a brigadier general. On November 25, Sherman Bell, the Adjutant General of Colorado wrote to Theodore Roosevelt, "Colonel Charles L. Cooper, our old Rough Rider stand-by at San Antonio... is in Denver." Bell requested that Cooper be assigned to the National Guard, a special assignment which Roosevelt heartily approved. Cooper finally retired to the home of his daughter, Florence, at 160 E. Granville Road, Worthington, Ohio. There he was visited shortly before his death in 1918 by former President Roosevelt.[248]

The first commander of Company H, Louis Henry Carpenter had a highly distinguished career in the U.S. Army. In 1887, General Sheridan chose Carpenter to establish Fort Myer outside of Washington, DC as the cavalry showplace of the nation. Carpenter's troops formed Sheridan's escort at the centennial celebration of the U. S. Constitution in Philadelphia, September 15-17, 1887. On July 1, 1888, the honor guard from Fort Myer was at Gettysburg for the twenty-fifth anniversary of the battle. Carpenter's old Civil War unit, the Sixth Cavalry held a reunion for survivors of the Battle of Fairfield on July 3. After a tour of the battle site and a campfire with over 500 in attendance, on July 4 they visited the Gettysburg battlefield. Sighting their old comrade in arms on guard with his command, the Fairfield veterans charged Major Carpenter's troopers. The amused veterans claimed that "for the first time, an old officer of the gallant old Sixth was completely surprised...He took his capture, however, philosophically, ...like an old trooper." After Carpenter convinced the "gallant Sixth" that he had been prevented by duty from attending the reunion, greetings were exchanged and the veterans continued on to the cemetery.[249]

On April 8, 1898, while Colonel Carpenter commanded Fort Sam Houston, he belatedly received the Medal of Honor for his service thirty years before in the relief of Beecher Island and the action at Beaver Creek. Carpenter retired to his home in Philadelphia on October 19,1899 as a brigadier general. For thirty-eight years, he had been a soldier. Carpenter continued an extensive work in family history begun by his father. In 1912, he published a complete genealogy entitled *Samuel Carpenter and His Descendants*. Carpenter died from heart problems at his home, 2318 DeLancey Place in Philadelphia on January 21, 1916 just short of his seventy-seventh birthday. He was buried at Sweedesboro, New Jersey.[250]

The troopers of Company H who had served under Carpenter, Cooper, Hunt, and others settled in every part of the United States and engaged in a variety of careers. Some married and had children, while others, after years of military life, remained single wanderers. Most viewed their military service with pride and a sense of accomplishment. They took every opportunity to reminisce about their experiences with members of their community and other veterans. Some of them including George Newman and Charles Terry, veterans of the capture of Mangus, applied for and received the Indian Wars Campaign Badge. Others joined veterans' organizations to renew their contacts with old military comrades. In addition to Reuben Waller, the subscribers' list for *Winners of the West*, a publication of the National Indian War Veterans association, included members of Troop H such as Louis Anderson, Benjamin Bard, Alfred Carter, Lewis Thomas and Simon Turner. For some veterans, military service had become a way of life, while for many of the others military service had provided the tools with which to build their later lives.[251]

Louis Anderson, born a slave in Fort Valley, Georgia, worked as a teamster following his enlistment. He settled in El Paso, Texas where in support of his pension application in 1922, neighbors asserted that "Louis is an old man who has the regard and respect of everyone in this community." He received a pension of $50 per month. Anderson died in El Paso on March 23, 1928 at the age of 81. Benjamin Bard married three times and had two sons and a daughter. He worked as a car sweeper for the Pennsylvania Railroad and the Baltimore & Ohio Railroad. Bard died at age 74 in Crestmont, Pennsylvania on August 19, 1927. Andrew Emory married Dora Packard in 1900 and had three sons and two daughters. He farmed and hired himself out as a plumber. Emory died on July 21, 1919 at Deer Creek, Minnesota. Douglas E. Lee served ten years and then returned to his home in Columbus, Ohio. He became a coal miner and worked for a time on a turnpike near Pittsburgh, Pennsylvania. By 1890, Lee's rheumatism became so severe that he "could not stand the work" and he applied for a military pension. William Hawkins settled in the Washington, DC area, working as a laborer for the Government Printing Office. He married Ivy Lee Johnson, a chairwoman at the Treasury Department. They had five daughters and four sons. In 1929, Hawkins served as part of an official delegation of veterans attending the inauguration of President Herbert Hoover.[252]

William Webb, still limping because of a wound in the thigh received from an Apache warrior in July 1879, married Mary Beard and had three children. They lived in Pleasantville, New Jersey where he was employed as a real estate agent. In his later years, Webb lived with his daughter, Mary Mitchell, dying on August 17, 1930. William Brent, a member of Asa Weaver's patrol that located Victorio in 1880, moved to Leavenworth, Kansas in the 1890's and married Frances Lewis. They settled into a house at the corner of Fifth and Maple and had one son. Jacob Watkins lived in Kansas City, Missouri, where he was employed as a coachman. His wife Fannie, took in washing and ironing. Watkins died September 2, 1899, with a disease of the kidneys and bladder. Webb Chatmoun, a coal miner for many years, died of bronchial pneumonia at age 77. His widow, Minerva, had no means of support and soon became dependent on neighbors for food and fuel. After one enlistment, Randall Blunt moved to San Francisco, married, and had twins, Randall K. and Thema K. He died there on April 5, 1922 at the home of his daughter Thema Simpson and was buried at San Francisco National Cemetery. Henry Walker received a disability discharge in 1885 at Fort Davis. He married Louise Brown later that year in Jeffersonville, Indiana. They had two sons and a daughter. Walker died of pneumonia on April 29, 1914 at the Military Home in Marion, Indiana. Isaac Jackson, former sergeant and company clerk, received a medical discharge in 1891 and settled in Albuquerque, New Mexico. He married Julia

Berry in 1901. Jackson worked as a porter, saloonkeeper and jailer. He collapsed on the streets of Albuquerque of cardiac arrest and died October 24, 1902.[253]

William Perry Battle, a member of the patrol that captured Mangus, married Ellen Anderson and settled in Pittsburgh, Pennsylvania. He served in the militia with Battery B, Pennsylvania Light Artillery. When the body of President William McKinley passed through Pittsburgh following his assassination, Battle was chosen to blow taps. Battle worked many years for the Bureau of Mines of Pennsylvania. He was an officer of the Knights of Pythias, a member of the Veterans of Foreign Wars, and an active member of the Euclid Avenue AME Church. Battle died January 17, 1917 at his home in Pittsburgh from acute lobar pneumonia. George Foster, who served with Battle, married divorcee Olivia Cane in El Paso, Texas. He died there on October 19, 1907. Solomon Boller, the blacksmith on the patrol that captured Mangus, married Fannie Thompkins in Gallup, New Mexico. During the 1890's, they lived in Albuquerque. Joseph Cammel, another Troop H soldier present at the capture of Mangus lived with the Bollers until his death on December 15, 1899. Occasionally, other old soldiers in need, like John T. Taylor stayed with Boller. After the passage of a bill to benefit Indian War veterans in 1927, Boller decided to apply for a pension. He received $40 a month, increased to $50 in 1931, to $55 in 1937, to $72 in 1938 because he needed regular care, and finally to $100 in 1944. On November 12, 1951, his wife Fannie died. One month later, on December 23, 1951, Solomon Boller passed away in Los Angeles, California at age 90.[254]

Another member of Troop H who took care of former soldiers in need was Charles I. Taylor, a veteran of the Spanish-American War. Born in North Carolina in 1872, Taylor studied at Clark College in Atlanta. After his military service, he became manager of the Indianapolis ABCs baseball team in the Negro League. Taylor played baseball as a pinch hitter until he reached his forties and was "regarded by some players as the finest manager in black baseball history." One common story concerning Taylor tells that once when an umpire called him safe after stealing third base, he brushed himself off and said "Ladies and gentlemen, I am an honest man. The umpire's decision is incorrect. I therefore declare myself out." Pitcher Arthur W. Hardy commented that "his players were tremendously loyal to him. The ABCs in their dress and general decorum, more nearly approximated the modern professional athlete than any other group that I ever saw in those days." Taylor died on March 2, 1922. Another long-time baseball player, George Osborne, returned to Fort Ethan Allen, Vermont with the Quartermaster Corps after his service in the cavalry. He married Vesta Monroe from New York in 1919. Osborne retired from the army in 1935, but continued to work on post as a civilian until 1954. He died of cardiac arrest on April 26, 1983 at the age of 98.[255]

George Osborne. Passed away at the age of 98.

By the 1930's, the reputation of Company H being always in the vanguard of every frontier campaign had become nothing more than a proud memory. The old soldiers in the troop were those who had served on the border against Pancho Villa, not the veterans of the Indian campaigns. In 1931, the Tenth Cavalry left Fort Huachuca, Arizona and the southwestern frontier for the last time. A short manuscript found in what appears to be the contents of a drawer from the Troop H orderly room at Fort Huachuca seems to accurately portray the feeling of many of the veteran buffalo soldiers. "From 1866 to 1898 the regiment was constantly in Indian campaigns; this service from Montana (where many lost fingers and toes from cold), to the Mexican border. They spent more time in the field than in garrison; five engagements with Cheyennes, ten with Apaches, thirteen with Comanches. Indians called them 'Buffalo soldiers' because they never stopped advancing." the writer continued, "For 65 years, the 10th has been in the habit of building their own homes, making the desert or jungle a fit place to live."[256]

The veterans' campfires slowly died away and the songs of the frontier troopers fell silent. Yet their legacy remains. Despite the constant hardships of long patrols and endless drills in all types of weather and terrain, despite the added hardship of senseless yet consistent prejudice, the Buffalo soldiers remained intensely loyal and committed to the military service they had freely chosen. In the first generations after slavery, military service provided Black men a chance to significantly improve their lives and lasting experiences that proved vital to the way in which they lived. "Always in the vanguard," the veterans of Company H, Tenth United States Cavalry provided an enviable example of consistent and dedicated service to the frontier army.

NOTES

[1] Eric J. Wittenberg, *Gettysburg's Forgotten Cavalry Actions* (Gettysburg, PA: Thomas Publications, 1998), 84-87; Theophilus F. Rodenbough and William L. Haskin, *The Army of the United States: Historical Sketches of Staff and Line with Portraits of Generals-in-Chief* (New York: Maynard, Merrill and Company, 1896), 236; Appointment, Commission and Promotion file, Louis H. Carpenter, RG 94, (National Archives).

[2] ACP file, Louis H. Carpenter, RG 94; Francis B. Heitman, *Historical Record and Dictionary of the United States Army From its Organization, September 29, 1789 to March 2, 1903* (Washington, DC: Government Printing Office, 1903) 284.

[3] Eric J. Wittenberg, The *Battle of Brandy Station: North America's Largest Cavalry Battle*, (Charleston, SC: The History Press, 2010), 100-154; Rodenbough and Haskin, *Army of the United States*, 235.

[4] ACP file, Louis H. Carpenter, RG 94; William H. Powell, *Records of Living Officers of the United States Army*, (Philadelphia, 1890) 111; *Official Army Register for 1900*, (Washington, DC: Adjutant General's Office) 344; San Antonio *Light*, May 3, 1898.

[5] ACP file, Louis H. Carpenter, RG 94; *OR*, Series 1, v 45, pt 1, 1197 and pt 2, 165-166.

[6] Reuben Waller may have been referring to Colonel Robert McCulloch of Cooper County, Missouri who commanded a brigade in Forrest's cavalry corps when he claims to have had a master who was a Confederate general. McCulloch's command was in all of the engagements that Waller mentions. Waller was obviously in Missouri rather than Kentucky by the time of the Border Ruffian Wars and there is some family tradition that he was sold to the man who took him to war.

[7] Reuben Waller, "History of a Slave Written by Himself at the Age of 89 Years," *The Battle of Beecher Island, Fought - September 17, 18, 1868*, (Wray, CO: Beecher Island Battle Memorial Association, revised printing, Sterling, CO: Royal Printing Company, 1985) 193-194.

[8] Ibid; Hutchinson *News Herald*, August 5, 1940.

[9] *Army-Navy Journal*, April 7, 1866, 525; Art T. Burton, *Black, Buckskin and Blue: African American Scouts and Soldiers on the Western Frontier*, (Austin, TX: Eakin Press, 1999) 130; ACP file, Louis H. Carpenter, RG 94.

[10] E.L.N. Glass, *The History of the Tenth Cavalry, 1866-1921*, (Fort Collins, CO: The Old Army Press, 1972) 13; *Army-Navy Journal*, April 28, 1866, 569.

[11] William H. and Shirley A. Leckie, *Unlikely Warriors: General Benjamin H. Grierson and His Family*, (Norman, OK: University of Oklahoma Press, 1984) 143-144; ACP file, Louis H. Carpenter, RG 94.

[12] *Army-Navy Journal*, May 4, 1867, 588; Glass, *History of the Tenth Cavalry*, 13 -14; San Antonio *Light*, May 3, 1898.

[13] William H. Leckie, *The Buffalo Soldiers: A Narrative of the Black Cavalry in the West*, (Norman, OK: University of Oklahoma Press, revised 2003), 14; Enlistment records for John Allen, Frank Bloodson, Henry Carpenter, Perry Curry, Dunn Day, Alfred Dixon, Robert Edwards, Daniel Grissom, Henry Harper, George Harris, John Homager, George Martin, Scott Pinkley, Frank Rogers, Charles Shavers, Amos Smith, John Thompson, Robert Thrash, Butler Tillman, Jerry Williams, Joseph Williams, RG 94 (National Archives).

[14] Military pension record for Jacob Young (National Archives); Harold Ray Sayre, *Warriors of Color*, (Fort Davis, TX, 1975) 139.

[15] Military pension records for James H. Clayton and James H. Thomas; Enlistment records for John Clark, Jacob Ewing, Ezariah S. Freeman, Richard Garrison, Thomas Haydon, Alfred McPherson, John D. Price, Charles Sampson, Sidney Sanders, Alexander Adams, Daniel Grisson and Mitchell Jones, RG 94 (National Archives).

[16] Military pension records for George Garnett, George Goldsby, Henry Allen and William Black; Enlistment record for George Garnett, RG 94 (National Archives).

[17] Enlistment records for John Claggett, Joseph Claggett, Samuel Jackson and Reuben Waller, RG 94 (National Archives); correspondence with Warren Robinson, great-grandson of Reuben Waller; Waller, "History of a Slave," 194.

[18] Enlistment Records for 1867, RG 94 (National Archives); United States Bureau of the Census, Ninth Census, 1870, Kansas.

[19] Muster roll for Company H, Tenth U.S. Cavalry, August 1867, U.S. Army Continental Commands, 1821-1920, RG 98 (National Archives); Glass, *The History of the Tenth*, 13-14; Leckie, *The Buffalo Soldiers,* 12-13, 51.

[20] Regimental Returns, Tenth Cavalry, July 1867, RG 94 (National Archives); Leckie, *The Buffalo Soldiers,* 14-15.

[21] Leckie & Leckie, *Unlikely Warriors,* 149-150; Burton, *Black, Buckskin and Blue,* 178-80.

[22] Muster roll for Company H, August 1867, RG 98 (National Archives); Waller, "Colored Troopers' Hair is too Short for Scalping," *Winners of the West*, Vol. 1 (July 1924).

[23] Muster rolls for Company H, August and October 1867, RG 98 (National Archives).

[24] Muster rolls for Company H, December 1867 and April 1868, RG 98 (National Archives); Leckie & Leckie, *Unlikely Warriors,* 147-148.

[25] Muster rolls for Company H, December 1867, February, April, June, August and October 1868, RG 98; Letter of Major John E. Yard to Adjutant, detachment near Fort Wallace, May 28, 1868, RG 98; Letter of Captain Henry C. Bankhead to Adjutant, Fort Wallace August 9, 1868, RG 98; Letter of Granville Lewis, Fifth Infantry to Adjutant Tenth Cavalry August 9, 1868, RG 393 (National Archives).

[26] *Army-Navy Journal*, April 11, 1868, 542; Muster rolls for Company H, June 1868, RG 98 (National Archives); Leckie & Leckie, *Unlikely Warriors*, 152-154.

[27] Muster rolls for Company H, June and August 1868, RG 98; Letter of Lt. Edward A. Belger to Commanding Officer detachment Tenth Cavalry July 14, 1868, RG 98 (National Archives).

[28] Muster rolls for Company H, August and October 1868, RG 98 (National Archives).

[29] Muster rolls for Company H, October 1868, RG 98 (National Archives); *Harper's Weekly*, September 7, 1867.

[30] Lincoln *Sentinel-Republican*, June 30, 1932, obituary of James J. Peate; *The Battle of Beecher Island, Fought - September 17, 18, 1868*, (Wray, CO: Beecher Island Battle Memorial Association, revised printing, Sterling, CO: Royal Printing Co, 1985); Clint E. Chambers and Paul H. Carlson, *Comanche Jack Stilwell: Army Scout and Plainsman,* (Norman: University of Oklahoma, 2019) 45-50.

[31] *The Battle of Beecher Island*, 96-100; Louis H. Carpenter, "The Story of a Rescue," *Winners of the West*, Vol. IX, No. 3 (February 1934); Muster roll of Company H, September-October 1868, RG 98 (National Archives).

[32] Lincoln *Sentinel-Republican*, June 30, 1932; Heitman, *Historical Record*, 423; Reuben Waller, "Trooper of the Tenth U.S. Cavalry Claims to be the First Man to the Rescue at Beecher Island," *Winners of the West*, Vol. II (August 1925); Merrill, Samuel, *The Seventieth Indiana Volunteer Infantry in the War of the Rebellion* (Indianapolis: The Bowen-Merrill Company, 1900), 110.

[33] *The Battle of Beecher Island*, 96-100; *Winners*, (February 1934); Muster roll of Company H, September-October 1868, RG 98 (National Archives).

[34] *The Battle of Beecher Island*, 96-100, 194-195; Reuben Waller, *Winners of the West*, Vol. 5, (August 1928); *Winners*, (February 1934).

[35] *Ibid.*

[36] *The Battle of Beecher Island*, 96-97, 194-195; Muster roll of Company H, September-October 1868, RG 98; Court-martial record for Ephraim Smith, Records of the Judge Advocate General, RG 153 (National Archives).

[37] Muster roll of Company H, September-October 1868, RG 98 (National Archives); *The Battle of Beecher Island*, 196; Cyrus Townsend Brady, *Indian Fights and Fighters*, (Lincoln, NE: University of Nebraska Press, 1971) 123-145.

[38] Muster roll of Company H, September-October 1868, RG 98 (National Archives); "The Fighting 10th Whipped the Indians at Beaver Creek," *Winners*, (October 1924); *Indian Fights and Fighters*, 125.

[39] *Indian Fights and Fighters*, 129-131; Muster roll of Company H, September-October 1868, RG 98 (National Archives).

[40] *Winners*, (October 1924); *Indian Fights and Fighters*, 132-133; ACP file for Carpenter, RG 94; Muster roll of Company H, September-October 1868, RG 98 (National Archives).

[41] *Indian Fights and Fighters*, 134; ACP file for Carpenter, RG 94; Muster roll of Company H, September-October 1868, RG 98 (National Archives).

[42] Leckie, *Buffalo Soldiers*, 39-40; Leckie & Leckie, *Unlikely Warriors,* 158-159.

[43] Muster roll of Company H, November-December 1868, RG 98; ACP file for Carpenter, RG 94; Carded Medical records for Silas Jones, Alfred Owings, Simon Peter and John Billings, RG 94 (National Archives).

[44] Muster roll of Company H, January-February 1869, RG 98 (National Archives); *Army-Navy Journal*, February 6, 1869, 386; Court martial records for John Savage, John Remington, Albert Horton, George Cobb, RG 153 (National Archives).

[45] *Army-Navy Journal*, March 24, 1869, 562; Letter of Bvt MGen. William B. Hazen, Medicine Bluff Creek Indian Reservation to MGen John M. Schofield, April 8, 1869, RG 98 (National Archives); Medical record for Alexander Adams, RG 94; Muster roll of Company H, May-June 1869, RG 98 (National Archives).

[46] Forrestine Cooper Hooker, *Child of the Fighting Tenth: On the Frontier with the Buffalo Soldiers*, (New York: Oxford University Press, 2003) ed. Steve Wilson, 54; Barbara E. Fisher, "Forrestine Cooper Hooker's Notes and Memoirs on Army Life in the West, 1871-1876" (unpublished thesis: University of Arizona, 1963), 36, 43-44; Letter of Col Anderson D. Nelson to Bvt BGen Chauncey McKeever, May 28, 1869, RG 98 (National Archives).

[47] Letter of Col Anderson D. Nelson to Bvt BGen Chauncey McKeever, May 28, 1869, RG 98; ACP file for Carpenter, RG 94 (National Archives); *The Battle of Beecher Island*, 197.

[48] Post returns for Camp Supply, June, July 1869, RG 98; Report 1348 for the relief of Charles Banzhof, 51 Congress, Senate (Washington, DC: 1890); Court martial records for Amos Cormack, RG 153 (National Archives).

[49] Post returns for Camp Supply, August and October 1869, RG 98; U.S. Army Register of Enlistments, 1798-1914 for Amos Cormack, William Oliver and Charles Shavers, RG94; Court martial records for Alfred Dixon, William Johnson, William Oliver, Charles Shavers and James Wright, RG 153 (National Archives).

[50] Post returns for Camp Supply, August, October, December 1869 and April 1870, RG 98; Muster roll for Company H, January-February 1870, RG 98; U.S. Army Register of Enlistments, 1798-1914 for Henry Allen, Robert Edwards, Lewis Hayes, John Henry, Louis Mack and Anderson Wilson; Court Martial records for Robert Edwards, Lewis Hayes, William Hamlin, John Henry and William Ross, RG 153 (National Archives); United States Bureau of the Census, Ninth Census, 1870, Texas.

[51] Muster roll for Company H, May-June 1870, RG 98; ACP file for Carpenter, RG 94 (National Archives); *Winners of the West*, (March 1930); Robert Carriker, *Fort Supply, Indian Territory: Frontier Outpost on the Plains*, (Norman, OK: University of Oklahoma Press, 1970); *The Battle of Beecher Island*, 197.

[52] ACP file for Carpenter, RG94 (National Archives); Leckie, *The Buffalo Soldiers*, 54-55.

[53] Letter of LtCol Anderson D. Nelson to AAG US Army, October 15, 1870, RG 98; Post return for Camp Supply, October 1870, RG 98; Muster roll for Company H, November-December 1870, RG 98 (National Archives); *The Battle of Beecher Island*, 197; *Engagements with Hostile Indians*, 30.

[54] ACP file for Carpenter, RG 94 (National Archives); *The Battle of Beecher Island*, 198; Leckie & Leckie, *Unlikely Warriors*, 186-188.

[55] Post return for Fort Sill, May 1871, RG 98; ACP file for Carpenter, RG 94 (National Archives); Leckie & Leckie, *Unlikely Warriors,* 188-190.

[56] Appointment, Commission and Promotion file, Alexander S. B. Keyes, RG 94 (National Archives); Thomas Harwood, *History of New Mexico Spanish and English Missions of the Methodist Episcopal Church From 1850-1910,* Vol. 1, (Albuquerque, NM: El Abogado Press, 1908) 84-90.

[57] ACP file for Keyes, RG 94 (National Archives); Harwood, *History of Spanish and English Missions,* 109, 117; *Army-Navy Journal,* September 18, 1897.

[58] Post returns for Fort Sill, August, September, October 1871, RG 98; Muster roll for Company H, September-October 1871, RG 98; ACP file for Carpenter, RG 94; Medical record for Reuben Waller, RG 94 (National Archives); Robert G. Carter, *On the Border with Mackenzie,* (New York: Antiquarian Press, Ltd., 1961) 142-143.

[59] *Engagements with Hostile Indians,* 34; Letter of LtCol Anderson D. Nelson to AAG, Department of the Missouri, April 26, 1872, RG 98; Post returns for Fort Gibson, September, October and November 1872, RG 98; Muster roll for Company H, November-December 1872, RG 98 (National Archives).

[60] Post returns for Fort Gibson, October and November 1872, RG 98; Muster roll for Company H, November-December 1872, RG 98 (National Archives).

[61] *Engagements with Hostile Indians,* 35; ACP file for Carpenter, RG 94; Enlistment record for John Claggett, RG 94; Post returns for Fort Sill, May, July and August 1873, RG 98 (National Archives).

[62] Muster roll for Company H, September-October 1873, RG 98; Post returns for Fort Sill, January, February, May, June, and July 1874, RG 98; ACP file for Carpenter, RG 94; Military Pension record for George Goldsby (National Archives); Fisher thesis, 95-96.

[63] J. Wright Mooar, "Frontier Experiences of J. Wright Mooar," *West Texas Historical Association Year Book,* Vol. IV, (June 1928) 91-92; James L. Haley, *The Buffalo War: The History of the Red River Uprising of 1874,* (Garden City, NY: Doubleday and Company, 1976).

[64] Carl C. Rister, "Early Accounts of Indian Depredations," *West Texas Historical Association Year Book,* Vol. II, (June 1926) 46-48; Leckie, *The Buffalo Soldiers,* 118.

[65] Muster roll for Company H, July-August 1874, RG 98; ACP file for Carpenter, RG 94 (National Archives); Haley, *The Buffalo War.*

[66] Col John Davidson, Report of action at the Wichita Agency, August 22, 1874, RG 98 (National Archives); Haley, *The Buffalo War.*

[67] Col John Davidson, Report of action at the Wichita Agency, August 22, 1874, RG 98; Muster roll for Company H, July-August 1874, RG 98; ACP file for Carpenter, RG 94 (National Archives); *Daily Alta California*, August 31, 1874.

[68] Muster roll for Company H, September-October 1874, RG 98; Post returns for Fort Sill, August and September 1874, RG 98 (National Archives); Carter, *On the Border with Mackenzie*, 475-476.

[69] Mackenzie to AAG, Department of Texas, September 19, 1874; Mackenzie's Expedition as Described by a Special Correspondent to the New York Herald, September 29, 1874 in Wallace, Ernest, ed., *Mackenzie's Official Correspondence, 1873-1879* (Lubbock, TX: West Texas Museum Association, 1967), 93, 112-115, 119-122.

[70] Mackenzie to AAG, Department of Texas, September 19, 1874; Mackenzie's Expedition as Described by a Special Correspondent to the New York Herald, September 29, 1874 in Wallace, ed., *Mackenzie's Official Correspondence, 1873-1879*, 93, 112-115, 119-122; *Engagements with Hostile Indians,* 42.

[71] Muster roll for Company H, September-October 1874, RG 98; ACP file for Carpenter, RG 94 (National Archives); *Engagements with Hostile Indians,* 42.

[72] Muster roll for Company H, November-December 1874, RG 98; letter of LtCol Davidson to Gen Christopher C. Auger, November 23, 1874, RG 98 (National Archives); *Engagements with Hostile Indians,* 43; Sayre, *Warriors of Color*, 107-108.

[73] Muster roll for Company H, November-December 1874, RG 98; Military Pension record for Benjamin Bard (National Archives); Sayre, *Warriors of Color*, 107-108.

[74] Post Returns for Fort Sill, February and March 1875, RG 98; ACP file for Carpenter, RG 94 (National Archives); *Engagements with Hostile Indians,* 46-47; Leckie & Leckie, *Unlikely Warriors,* 219.

[75] Post Returns for Fort Sill, February and March 1875, RG 98; ACP file for Carpenter, RG 94 (National Archives); *Engagements with Hostile Indians,* 46-47; Leckie & Leckie, *Unlikely Warriors,* 219; Leckie, *The Buffalo Soldiers*, 136-139.

[76] United States Bureau of the Census, Tenth Census, 1880, Texas.

[77] Post Return for Fort Sill, March 1875, RG 98 (National Archives).

[78] Muster roll for Company H, Tenth U.S. Cavalry, April 30, 1875, U.S. Army Continental Commands, 1821-1920, RG 98; Appointment, Commission and Promotion file, "Military Record of Brigadier General Louis Henry Carpenter," 17, RG 94; Medical History of Fort Davis, "Description of Post," RG 94 (National Archives).

[79] Sayre, *Warriors of Color*, 108; Return for the Tenth Regiment of Cavalry, October 1875, RG 98; Appointment, Commission and Promotion file, Charles G. Ayres, RG 94, (National Archives).

[80] Document file for "Fort Davis – Subposts," Fort Davis National Historic Site, TX; Return for the Tenth Regiment of Cavalry, October 1875, RG 98; ACP file, Carpenter, RG 94 (National Archives) *Roster of Non-Commissioned Officers of the Tenth U.S. Cavalry* (1897; rpt. Bryan, Texas: J. M. Carroll and Co., 1983), 32.

[81] Post Returns for February, March, May, June and July 1876, Fort Davis, TX; ACP file, Carpenter, RG 94; ACP file, William R. Harmon, RG 94 (National Archives).

[82] Post Return for June 1877, Fort Davis, TX; Hostile, 75; Sayre, *Warriors of Color*, 108; Schubert, *Voices of the Buffalo Soldiers,* 147-148.

[83] Medical Records, Fort Davis, 1875-1885, RG 98 (National Archives); Sayre, *Warriors of Color*.

[84] *San Antonio Express*, August 22, 1877; *Record of Engagements with Hostile Indians within the Military Division of the Missouri from 1868 to 1882*, (Washington: Government Printing Office, 1882) 78.

[85] C. L. Sonnichsen, *The El Paso Salt War*, (El Paso, TX: Texas Western Press, 1961) 16, 32-33, 36-38, 48-58, 60; Post returns for December 1877 and January 1878, Fort Davis, TX; ACP file, Carpenter, RG 94 (National Archives).

[86] Telegram from Major George Andrews to Adjutant General, Department of Texas, December 20, 1877, RG 98; Post returns for December 1877 and January 1878, Fort Davis, TX, RG 98 (National Archives).

[87] *Engagements with Hostile Indians,* 76-77; General Orders #2, Department of Texas, January 22, 1878, RG 98; Telegram from Major George Andrews to Adjutant General, Department of Texas, December 27, 1877, RG 98; Letter from Lieutenant Robert G. Smither to Commanding Officer, Company C, March 28, 1878, District of the Pecos, 1878-1881, RG 393 (National Archives).

[88] Report of Operations, Company H, Tenth Cavalry from May 20 to August 29, 1878, Captain L. H. Carpenter; Special Order #69, May 16, 1878, RG 393; letter

of Colonel Benjamin Grierson to Major George L Andrews, May 1, 1878 and letter of First Lieutenant Robert G. Smither, Acting AAG, Tenth Cavalry to Captain Carpenter, June 4, 1878, files at Fort Davis National Historic Site, TX.

[89] Ibid.

[90] Report of Operations, Company H, Tenth Cavalry from May 20 to August 29, 1878, Captain L. H. Carpenter, files at Fort Davis National Historic Site, TX; Chambers and Carlson, Comanche Jack Stilwell, 118-119.

[91] Ibid.; Letter from Lieutenant Smither to Captain Carpenter, June 4, 1878, RG 393 (National Archives).

[92] Marcus E. Kinevan, *Frontier Cavalryman: Lieutenant John Bigelow with the Buffalo Soldiers in Texas*, (El Paso, TX: Texas Western Press, 1998), 148.

[93] Post Returns for April, May, June, and August 1879, Fort Davis, TX; Muster Rolls for Company H, April-September 1879, RG 98; Military Pension record for William Webb (National Archives); Sayre, *Warriors of Color*, 137.

[94] Report of Operations of Company H from October 23 to November 30, 1879, Report of Colonel Grierson to Assistant Adjutant General, Department of Texas, December 31, 1879 and Telegram from Major Napoleon B. McLaughlin, District of the Pecos Records, RG 393 (National Archives).

[95] Dan L. Thrapp, *Victorio and the Mimbres Apaches* (Norman and London: University of Oklahoma Press, 1974), 3-5, 188-193, 214-219; *New York Times*, November 28, 1880.

[96] Colonel Grierson to AAG, Department of Texas, May 21, 1880, RG 393 (National Archives).

[97] Ibid.; Report of Captain Louis H. Carpenter to AAG, Department of Texas, December 9, 1880, District of the Pecos, RG 393 (National Archives).

[98] Lieutenant Swisher to Company M, Tenth Cavalry, May 25, 1880, RG 393; Report of Captain Louis H. Carpenter to AAG, Department of Texas, December 9, 1880, RG 393; Special Orders #5, March 2, 1880, District of the Pecos, RG 393 (National Archives).

[99] Report of Colonel Grierson to AAG, Department of Texas, December 31, 1880, RG 393; Special Orders #17, District of the Pecos, June 27, 1880, RG 393 (National Archives).

[100] General Orders #3, District of the Pecos, July 9, 1880, RG 393; Special Orders #27, District of the Pecos, August 25, 1880, RG 393; Colonel Grierson to

Commanding Officer, Fort Bliss, July 18, 1880, RG 393; Telegram from Colonel Adolfo Valle to Colonel Grierson, July 25, 1880, RG 393 (National Archives).

[101] Report of Colonel Grierson to AAG, Department of Texas, September 20, 1880, District of the Pecos, RG 393 (National Archives); Robert K. Grierson, "Journal Kept on the Victorio Campaign in 1880" (typescript in Fort Concho Library and Archives).

[102] Ibid.; Colonel Grierson to Captain Carpenter, July 30, 1880, RG 393 (National Archives).

[103] Post Return for July 1880, Fort Davis, TX, RG 98; Letter from Captain Carpenter to Lieutenant Smither, June 20, 1880, Report of Captain Carpenter, December 9, 1880 and letter from Colonel Grierson to Captain Carpenter, July 30, 1880, RG 393, (National Archives).

[104] Report of Captain Carpenter, December 9, 1880, RG 393, (National Archives); Muster rolls for Company H, July-August 1880, RG 98; Military Pension records for Asa Weaver and William Brent (National Archives); *Roster of Non-commissioned Officers of the Tenth U.S. Cavalry* (Bryan, TX and Mattituck, NY: J.M. Carroll and Company, 1897), 28.

[105] Report of Operations, Company H, Tenth Cavalry from May 20 to August 29, 1878, Captain L. H. Carpenter, files at Fort Davis National Historic Site, TX; Report of Captain Carpenter, December 9, 1880, RG 393, (National Archives).

[106] "Military Record of Brigadier General Carpenter," RG 94, 20; Report of Captain Carpenter, December 9, 1880, District of the Pecos, RG 393; Muster rolls for Company H, July-August 1880, RG 98; Returns for the Tenth Regiment of Cavalry, June-August 1880, RG 98, (National Archives).

[107] Report of Colonel Grierson, September 20, 1880 and Report of Captain Carpenter, December 9, 1880, RG 393; Muster Roll for Company H, Tenth Cavalry, May-June 1880, RG 98 (National Archives); *Winners of the West*, Vol. 11 (April 1934).

[108] Sayre, *Warriors of Color*, 109; Report of Captain Carpenter, December 9, 1880, RG 393 (National Archives); *Federal Writers' Project: Slave Narratives*, vol. 16 (Washington, DC, 1941), interview with Madison Bruin, 169-173.

[109] Letter from Captain Carpenter to Lieutenant William Beck, August 31, 1880; letter from Lieutenant Beck to Captain Carpenter, September 8, 1880, letter from Captain Carpenter to Captain Nicholas Nolan, September 12, 1880, RG 393; Military Pension record for Asa Weaver (National Archives); Sayre, *Warriors of Color*, 406-409.

[110] Letter from Captain Carpenter to Lieutenant Beck, September 25, 1880 and telegram from Joachim Terrazas to Colonel George P. Buell, October 18, 1880, RG 393; Post Return for October 1880, Fort Davis, TX, RG 98 (National Archives).

[111] George W. Baylor, "The Last Fight on Texas Soil Between the Apaches and Texas Rangers," The Texas Rangers Association, 5 (December, 1905), 18-19; San Antonio *Express*, March 28, 1916; Waller, "Colonel George Wythe Baylor," 35.

[112] General Order 1, District of the Pecos, February 7, 1881, RG 393; Post Return for November 1880, Fort Davis, TX, RG 98 (National Archives).

[113] Post Returns for April, July, August, September, October 1881, January 1882 and August 1883, Fort Davis, TX, RG 98; Muster rolls for Company H, July-August and September-October 1881, RG 98 (National Archives).

[114] ACP file for Carpenter, RG 94; Muster Rolls for Company H, May-June and November-December 1882, RG 98 (National Archives); Sayre, *Warriors of Color,* 8, 12.

[115] Appointment, Commission and Promotion file for Charles L. Cooper, RG 94 (National Archives); Barbara E. Fisher, "Forrestine Cooper Hooker's Notes and Memoirs on Army Life in the West, 1871-1876" (unpublished thesis: University of Arizona, 1963), 4-6; Hooker, *Child of the Fighting Tenth*, 9, 20-21.

[116] ACP file for Charles L. Cooper, RG 94 (National Archives); Fisher thesis, 8-10.

[117] *Ibid.*

[118] Paul H. Carlson, *The Buffalo Soldier Tragedy of 1877* (College Station, TX: Texas A & M University Press, 2003), 69, 99, 105,133; Joseph H. King, "Brief Account of the Sufferings of a Detachment of United States Cavalry from Deprivation of Water," *The American Journal of the Medical Sciences*, 75 (1878), 405-408.

[119] ACP file for Charles L. Cooper, RG 94 (National Archives); Carlson, *Buffalo Soldier Tragedy*, 135.

[120] ACP file for Carpenter, "Military Record of Brigadier General Carpenter," 21-22, RG 94; Post returns for May and June 1882, Fort Davis, TX, RG 98; Muster Rolls for Troop H, Tenth Cavalry, October 1882 and February 1884, RG 98 (National Archives); Hooker, *Child of the Fighting Tenth*, 187, 191.

[121] Heitman, *Historical Register*, 1011; Biographical file on Francis H. Weaver, Fort Concho Archives; Chaplain's Report for April 1881 and May 1883.

[122] Thirty-Ninth Congress, Session 1, July 28, 1866, An Act to increase and fix the Military Peace Establishment for the United States, Section 27; G. B. Quackenboe, *Illustrated School History of the United States* (New York: D. Appleton and Company, 1868), 41-42; S. G. Goodrich, *A Pictorial History of the United States* (New York: Huntington and Savage, Mason and Law, 1860), 312.

[123] *Army and Navy Journal*, January 19, 1884; Hooker, *Child of the Fighting Tenth*, 166.

[124] U.S. Army Register of Enlistments, 1798-1914 for Thomas Maddox, RG94; Registers of Deaths in the Regular Army, 1860-1889 for Thomas Maddox and Benjamin Banks, RG94; Sayre, *Warriors of Color*, 83-103.

[125] Post return, April 1885, Fort Davis, TX, RG 98 (National Archives); Sayre, *Warriors of Color*, 138, 328.

[126] Post Returns for Fort Riley, January and February 1868, RG 98; Post Returns for Fort Sill, July 1871 and May 1874, RG 98; Post returns for Fort Davis, May and June 1882, RG 98 (National Archives); Omaha *Bee*, September 18, 1883.

[127] Sayre, *Warriors of Color*, 48, 83-85; Hooker, *Child of the Fighting Tenth: On the Frontier with the Buffalo Soldiers*, 56-57; *Army-Navy Journal*, February 26, 1870, 431; Winners of the West, Vol 14, (July 1937).

[128] *Army-Navy Journal*, September 7, 1867, 42; Fisher thesis, 37; Sayre, *Warriors of Color*, 62-75.

[129] Ninth U.S. Census, 1870, Indian Territory; Tenth U.S. Census, 1880, Texas, Jeff Davis County; Record of Enlistments, RG 94 (National Archives).

[130] George M. Mullins, Chaplain's Report for Fort Davis, 1877, RG 94; Report of the Secretary of War, 1889 (National Archives); Kinevan, *Frontier Cavalryman*, 148; Fisher thesis, 38, 59-60.

[131] Omaha *Bee*, April 1, 1883.

[132] Schubert, *On the Trail of the Buffalo Soldier II*, 99; Wooster, *Soldiers, Sutlers and Settlers*, 64-70; Carded Medical records for George W. Foster and John Morgan, RG 94 (National Archives).

[133] Carded Medical records for James Andrews, Charles Barnum, Dorsey Johnson and Henry Walker, RG 94; Military Pension records for Washington Hardaway and William Webb (National Archives); Sayre, *Warriors of Color*, 461.

[134] Carded Medical records for James Jackson and Colonel Miller, RG 94; Military Pension records for Charles Black, William Hawkins and William Webb (National Archives).

[135] Carded Medical records for Benjamin Banks, Charles Black and James H. Thomas, RG 94; Medical History of Post for Fort Concho and Fort Rice, RG 94 (National Archives).

[136] Carded Medical records for William C. Alexander, Joseph Claggett, Silas Jones, James A. Hill, James H. Thomas and others, RG 94 (National Archives).

[137] Carded Medical records for Dorsey Johnson, George Johnson, James Johnson, John Lisby and John Muchs, RG 94; Military pension record for Dorsey Johnson (National Archives).

[138] Court martial record for Alexander Brown, RG 153 (National Archives).

[139] Court martial records for Anderson Wilson, RG 153 (National Archives).

[140] Sayre, *Warriors of Color*, 396-399.

[141] Court martial records for John H. Curtiss, William Epps and Doc Mocfield, RG 153 (National Archives).

[142] Court martial records for Greenfer Shanklin and Benjamin Harris, RG 153 (National Archives).

[143] Court martial records for William Chase, Michael Finnegan and Charles Gray, RG 153 (National Archives).

[144] Court martial records for James Clayton, Amos Cormack and Henry Allen, RG 153 (National Archives).

[145] Court martial records for Henry Allen, RG 153 (National Archives).

[146] Court martial records for Silas Jones, Pollard Cole, William Hamlin and George Bumferts, RG 153 (National Archives).

[147] Court martial records for Robert Ellis, RG 153 (National Archives).

[148] Court martial records for Ephraim Smith, Dickson Hunter and Charles Jefferson, RG 153 (National Archives).

[149] Court martial records for John Henry, William Pierce and William Clark, RG 153 (National Archives).

[150] Court martial records for Ellson Clark and William Booter, RG 153 (National Archives); Arthur Murray, *A Manual for Courts-Martial*, (New York: John Wiley and Sons, 1893), 71.

[151] Court martial records for George Douglass, Andy Clayton and Louis Bell, RG 153 (National Archives).

[152] Court martial records for Robert Valentine and Charles Reed, RG 153 (National Archives).

[153] Court martial records for Samuel Porter and George Horton, RG 153 (National Archives); Sayre, *Warriors of Color*, 158-162.

[154] Court martial record for Lewis Hayes, RG 153 (National Archives).
[155] Muster roll for Troop H, Tenth Cavalry, May-June 1885, RG 98; Returns of the Tenth Cavalry, April, May and June 1885, RG 98, (National Archives); Forrestine C. Hooker, *When Geronimo Rode*, (Garden City, NY: Doubleday, Page and Co., 1924) 25.

[156] Frederic Remington, *Selected Writings*, compiled by Frank Oppel, (Secaucus, NJ: Castle, 1981) 175-176, 183.

[157] Sayre, *Warriors of Color*, 47, 55, 109; Hooker, *Child of the Fighting Tenth*, 187; Muster roll for Troop H, November-December 1885, RG 98 (National Archives).

[158] Returns of the Tenth Cavalry, January, April and May 1886, RG 98 (National Archives); Omaha *Bee*, April 1, 1883; Leckie and Leckie, *Unlikely Warriors*, 286-289; Hooker, *Child of the Fighting Tenth*, 165.

[159] Muster roll for Troop H, September-October 1886, RG 98; Returns of the Tenth Cavalry, December 1886, RG 98 (National Archives).

[160] ACP file for Charles L. Cooper, RG 94; Report of Scout, Captain Charles Cooper, October 20, 1886, RG 98; Muster roll for Troop H, September-October 1886, RG 98 (National Archives); Sayre, *Warriors of Color*, 110-111.

[161] *Ibid*; Hooker, *Child of the Fighting Tenth*, 234; Hooker, *When Geronimo Rode*, 284-287.

[162] Sayre, *Warriors of Color*, 36-37, 111; Returns of the Tenth Cavalry, December 1886 and July 1887, RG 98 (National Archives); Hooker, *Child of the Fighting Tenth*, 237.

[163] Military pension record for Colonel E. Miller; Return of the Tenth Cavalry, December 1888, RG 98 (National Archives).

[164] Glass, *History of the Tenth*, 28.

[165] Muster roll for Troop H, January-February 1890, RG 98; Return of the Tenth Cavalry, June 1890, RG 98; Court-martial record for Louis Bell, RG 153 (National Archives).

[166] Muster rolls for Troop H, January-February 1891, March-April 1892, RG 98; Return of the Tenth Cavalry, May 1892, RG 98; ACP file for Thaddeus W. Jones, RG 94 (National Archives); Heitman, *Historical Record*, 582.

[167] Returns of the Tenth Cavalry, May 1892, July and September 1893, RG 98; Muster roll for Troop H, July-August 1894, RG 98 (National Archives).

[168] U.S. Army Register of Enlistments, 1798-1914 for 1893-1898, RG 94; Returns of the Tenth Cavalry for 1893- 1896, RG 98 (National Archives).

[169] Appointment, Commission and Promotion file for Levi P. Hunt, RG 94 (National Archives); Heitman, *Historical Record*, 556; file on Lieutenant Levi P. Hunt, Fort Davis National Historic Site, TX; Hooker, *Child of the Fighting Tenth*, 51, 119, 179.

[170] Returns of the Tenth Cavalry, 1893 - 1894, March, April, May, June, July and August 1895, RG 98 (National Archives).

[171] Muster roll of Troop H, September-October 1895, RG 98; Return of the Tenth Cavalry, October 1895, RG 98; Special Order 61, 1895, Tenth U.S. Cavalry, RG 98 (National Archives).

[172] General Order 1, Tenth U.S. Cavalry, June 7, 1897; General Order 2, July 7, 1897; General Order 4, August 3, 1897, RG 98 (National Archives).

[173] Muster roll for Troop H, July-August 1897, RG 98; Returns of the Tenth Cavalry, August and September 1897, RG 98 (National Archives); Sayre, *Warriors of Color*, 327.

[174] Muster roll for Troop H, September-October 1897, RG 98 (National Archives); Remington, *Selected Writings*, 7.

[175] Remington, *Selected Writings*, 7, 178.

[176] Richmond *Planet*, January 29, 1898; Frederic Remington, *Selected Writings*, compiled by Frank Oppel, (Secaucus, NJ: Castle, 1981), x.

[177] Richmond *Planet*, February, 19, 1898.

[178] Returns of the Tenth Cavalry, January and February 1898, U.S. Army, Continental Commands, 1821-1920, RG 98; Enlistment Record for Charles Faulkner, RG 94 (National Archives).

[179] Return of the Tenth Cavalry, March 1898, RG 98 (National Archives).

[180] Theophilus G. Steward, *The Colored Regulars in the United States Army* (New York: Arno Press and the New York Times, 1969), 255-256.

[181] Ibid., 98, 256; Herschel V. Cashin, *Under Fire with the Tenth U. S. Cavalry* (Niwot, CO: University Press of Colorado, 1993), 217; Richmond *Planet*, May 14, 1898.

[182] Returns of the Tenth Cavalry, June and July 1898, RG 98 (National Archives); Richmond *Planet,* July 23, 1898; Illinois *Record*, August 13, 1898.

[183] Sayre, *Warriors of Color*, 43; Schubert, *On the Trail of the Buffalo Soldier II*, 43; ACP file for Charles Cooper, RG 94 (National Archives); New York *Evening Sun*, April 25, 1898.

[184] ACP files for Louis Carpenter and Alexander Keyes, RG 94 (National Archives); Heitman, *Historical Record*, 284; San Antonio *Light*, April 16, April 19 and May 3, 1898.

[185] Remington, *Selected Writings*, 331; Cashin, *Under Fire*, 225, 258.

[186] Cashin, *Under Fire*, 218-219, 225; Steward, *The Colored Regulars*, 149.

[187] Cashin, *Under Fire*, 226-227; Steward, *The Colored Regulars*, 263.

[188] Cashin, *Under Fire*, 96, 153-157, 226-228; Remington, *Selected Writings*, 343; Richmond *Planet*, July 16, 1898.

[189] Richmond *Planet*, July 23, 1898; Illinois *Record*, August 13, 1898; Cashin, *Under Fire*, 220.

[190] Frank N. Schubert, *Black Valor, Buffalo Soldiers and the Medal of Honor, 1870-1898*, (Wilmington, DE: Scholarly Resources, Inc, 1997) 135-139; Cashin, *Under Fire*, 176, 347; Muster roll for Troop H, June 1898, RG 98 (National Archives).

[191] *Ibid.*

[192] Medal of Honor file for Dennis Bell, RG 94 (National Archives); Schubert, *Black Valor*, 135-139.

[193] ACP file for Charles G. Ayres, RG 94 (National Archives); Cashin, *Under Fire*, 203; Frank N. Schubert, *On the Trail of the Buffalo Soldier: African Americans in the US Army*, (Wilmington, DE: Scholarly Resources Inc, 1995) 13, 419; *Washington Bee*, November 12, 1898.

[194] Richmond *Planet*, November 19, 1898; Return of the Tenth Cavalry, October 1898, RG 98 (National Archives).

[195] Richmond *Planet*, November 19, November 26, December 17, 1898; Illinois *Record*, December 3, 1898; Muster roll for Troop H, November-December 1898, RG 98 (National Archives).

[196] Richmond *Planet*, November 26, December 13 and 17, 1898; Muster roll for Troop H, November-December 1898, RG 98 (National Archives).

[197] Richmond *Planet*, December 13, 1898.

[198] Muster rolls for Troop H, July-August 1899 and January-February 1900, RG 98; Returns of the Tenth Cavalry, January and April 1899, January 1900, RG 98 (National Archives).

[199] Medal of Honor file for Dennis Bell, RG 94 (National Archives); Schubert, *Black Valor*, 140-141.

[200] ACP file for Louis H. Carpenter, RG 94 (National Archives); *Official Army Register for 1900*, 344, 365.

[201] Returns of the Tenth Cavalry, January and February 1900, RG 98 (National Archives); United States Bureau of the Census, Twelfth Census, 1900, Texas and Wyoming.

[202] Returns of the Tenth Cavalry, November 1900, April and May 1901, RG 98 (National Archives); Glass, *History of the Tenth*, 50.

[203] Returns of the Tenth Cavalry, May, June and July 1901, RG 98; Return of the Second Squadron, Tenth Cavalry, July 1901, RG 98 (National Archives); Glass, *History of the Tenth*, 50.

[204] Returns of the Tenth Cavalry, August, September and December 1901, January, February and March 1902, RG 98 (National Archives).

[205] Returns of the Tenth Cavalry, July and August 1902, RG 98 (National Archives); United States Bureau of the Census, Thirteenth Census, 1910, Wyoming; Frank N. Schubert, "Black Soldiers on the White Frontier: Some Factors influencing Race Relation," *Phylon* 32 (1971), 411-413.

[206] Returns of the Tenth Cavalry, June 1902, January and February 1903, RG 98; ACP file for Charles T. Boyd, RG 94 (National Archives).

[207] Returns of the Tenth Cavalry, December 1903 and May 1904, RG 98; Robert Peterson, *Only the Ball was White: A History of Legendary Black Players and All-Black Professional Teams* (Oxford University Press, 1992), 251-252.

[208] Returns of the Tenth Cavalry, October 1906 and March 1907, RG 98; ACP files for Charles T. Boyd and James S. Greene, RG 94 (National Archives); Report of the Secretary of War, Fifty-eighth Congress (Washington, DC: Government Printing Office, 1903).

[209] Returns of the Tenth Cavalry, April, September and October 1907, June and July 1908, RG 98; ACP files for Charles T. Boyd, Eugene P. Jervey, Jr. and James S. Greene, RG 94 (National Archives); Glass, *History of the Tenth*, 50 - 55.

[210] Returns of the Tenth Cavalry, February, October and November 1908 and April and May 1909, RG 98 (National Archives); Glass, *History of the Tenth*, 50 - 52.

[211] Enlistment Record for George Coleman, RG 94 (National Archives); Glass, *History of the Tenth*, 40; New York *Daily Tribune*, July 27, 1909, p4.

[212] David Work, "The Buffalo Soldiers in Vermont, 1909 – 1913," *Vermont History,* 73 (Winter/Spring 2005): 63–75; Glass, *History of the Tenth*, 59; *Burlington Free Press*, December 22, 1972.

[213] United States Bureau of the Census, Twelfth and Thirteenth Census, 1900 and 1910, Vermont; Work, "The Buffalo Soldiers in Vermont," 65-68; *Rutland Daily Herald,* July 31, August 4 and August 14 1909.

[214] United States Bureau of the Census, Thirteenth Census, 1910, Vermont; Work, "The Buffalo Soldiers in Vermont," 64; Enlistment Record for William Thacker, RG 94 (National Archives).

[215] Glass, *History of the Tenth*, 59, 63; Work, "The Buffalo Soldiers in Vermont," 70; Marvin Fletcher, *The Black Soldier and Officer in the United States Army, 1891–1917*(Columbia: University of Missouri Press, 1974), 119–123.

[216] Glass, *History of the Tenth*, 63; Work, "The Buffalo Soldier in Vermont," 69-70; John Buechler,*"Buffalo Soldiers in the Green Mountains,"* Chittenden *County Historical Society Bulletin*, 5:3 (April 1970).

[217] Enlistment Record for Matthew Carlisle, RG 94 (National Archives); *Barre Daily Times*, October 11, 1911; *Bennington Evening Banner*, October 11, 1911.

[218] *The Norwalk Hour*, Oct. 11, 1911; *Barre Daily Times*, October 11, 1911 and February 29, 1912; *Bennington Evening Banner*, October 11, 1911; Work, "The Buffalo Soldier in Vermont," 70.

[219] Returns of the Tenth Cavalry, July, September and October 1913, RG 98 (National Archives); Work, "The Buffalo Soldier in Vermont," 71-73.

[220] Glass, *History of the Tenth*, 63-64; *Burlington Free Press*, December 13, 1913; Interview with Major General John B. Brooks, Fort Huachuca Museum files.

[221] Glass, History of the Tenth, 64-65; David K. Work, "Enforcing Neutrality: The Tenth U.S. Cavalry on the Mexican Border, 1913-1919," Western Historical Quarterly, 40 (Summer, 2009), 182-186.

[222] Ibid.; Returns of the Tenth Cavalry, September and December 1914, RG 98 (National Archives).

[223] Orders of Troop H, Tenth Cavalry, Troop Order 3, February 1, 1915, RG 98; Troop Order 4, February 14, 1915, RG 98 (National Archives); Glass, *History of the Tenth*, 65; Work, "Enforcing Neutrality," 184.

[224] Glass, *History of the Tenth*, 75-76, 135; Michael Lee Lanning, *The African American Soldier: From Crispus Attucks to Colin Powell*, (Secaucus, NJ, 1997), 117-119.

[225] Return of the Tenth Cavalry, April 1916, RG 98 (National Archives); Glass, *History of the Tenth*, 138-139; John Buechler, "Buffalo Soldiers in the Green Mountains," *Chittendon County Historical Society Bulletin*, Vol. 5, November, 1969 and April 1970.

[226] Glass, *History of the Tenth*, 81; Lanning, *The African American Soldier*, 117-119.

[227] John Minton, *The Houston Riot and Courts-Martial of 1917* (Munguia Printers: San Antonio, TX, undated) 3,12-14, 16, 28.

[228] Jesse Jackson, Jr., *A Social History of the Tenth Cavalry, 1931-1941*, (Fort Leavenworth, KS: US Army Command and General Staff College, 1976); Enlistment Records for Pinkney Armstrong and Clifford Sandridge, RG 94 (National Archives).

[229] Colonel Harold B. Wharfield, *With Scouts and Cavalry at Fort Apache*, ed. John Alexander Carroll, (Tucson, AZ: Arizona Pioneers' Historical Society, 1965), 1-3.

[230] Colonel Harold B. Wharfield, *With Scouts and Cavalry at Fort Apache*, ed. John Alexander Carroll, (Tucson, AZ: Arizona Pioneers' Historical Society, 1965), 1-3; Wharfield, *10th Cavalry and Border Fights*, (El Cajon, CA, 1965).

[231] James P. Finley, "The Buffalo Soldiers at Fort Huachuca," *Fort Huachuca Illustrated*, Volume 2, (1996), 58; Cornelius C. Smith, *Fort Huachuca:* The *History of a Frontier Post* (Fort Huachuca Museum, 1977), 229-31; *The Arizona Republican*, July 4, 1919; *Bisbee Daily Review*, July 4, 1919; Interview with Major General John B. Brooks, Fort Huachuca Museum files.

[232] United States Bureau of the Census, Fourteenth Census, 1920, Arizona; Enlistment Records for Atwood Mitchell, Garfield Dyas and George A. Lee, RG 94 (National Archives).

[233] Smith, *Fort Huachuca:* The *History of a Frontier Post,* 230-231; Interview with Major General Brooks.

[234] Glass, *History of the Tenth*, 86-87; Finley, *Fort Huachuca Illustrated*, Volume 2 (1996):58.

[235] Jackson, *A Social History of the Tenth Cavalry*, 25; David K. Work, "The Fighting Tenth Cavalry: Black Soldiers in the United States Army 1892-1918," (unpublished thesis: Oklahoma State University, 1998), 130.

[236] Jackson, *A Social History of the Tenth Cavalry*, 29; Work, "The Fighting Tenth Cavalry," 131; Finley, "The Buffalo Soldiers at Fort Huachuca," *Fort Huachuca Illustrated*, Volume 2, (1996), 67; Interview with Major General Brooks.

[237] Lanning, *The African American Soldier,* 154-155, 177; Jackson, *A Social History of the Tenth Cavalry*, 63-64; Work, "The Fighting Tenth Cavalry," 131.

[238] Reuben Waller, "History of a Slave," 198-199; Reuben Waller, "Trooper of the Tenth U.S. Cavalry Claims to be the First Man to the Rescue at Beecher Island," *Winners of the West*, Vol. II (August 1925), "A Survivor of the Beecher Island Fight of Sixty Years Ago," *Winners,* Vol. VI (May 1929); Lincoln *Sentinel-Republican*, June 30, 1932, obituary of James J. Peate.

[239] *Winners of the West*, Vol. 1 (July 1924); telephone conversations and correspondence with Warren Robinson, great-grandson of Reuben Waller.

[240] Sayre, *Warriors of Color*, 139-143, 327; Military pension record for Philip Jones.

[241] Sayre, *Warriors of Color*., 5-7, 11, 17, 21-22; Military pension records for Aaron Archer and Alexander Brown (National Archives)..

[242] Military pension records for Washington Hardaway and Charles Black (National Archives).

[243] Military pension records for Colonel Miller and Asa Weaver (National Archives).

[244] Military pension record for George Goldsby (National Archives); Leckie, *The Buffalo Soldiers*, 164.

[245] *Ibid.*

[246] Military pension record for Michael Finnegan (National Archives).

[247] Military pension records for Jacob Young and George Garnett (National Archives); Sayre, *Warriors of Color*, 110.

[248] ACP files for Levi Hunt, Thaddeus Jones and Charles Cooper, RG 94 (National Archives); Heitman, *Historical Record*, 556

[249] ACP file for Louis Carpenter, RG 94 (National Archives); "Proceedings of the Fifth Annual Reunion of the Survivors of the Sixth US Cavalry," July 3, 1888, Heinrich G. Mueller, Secretary.

[250] ACP file for Louis Carpenter, RG 94 (National Archives); *Official Army Register for 1900*, 344, 365; Edward Carpenter and Louis Henry Carpenter, *Samuel Carpenter and His Descendants* (Philadelphia: J.B. Lippincott Company, 1912), 120, 127.

[251] *Winners of the West*, Vol. IV (March 1927); *Winners*, Vol. VII (November 1930); Schubert, *On the Trail of the Buffalo Soldier II*, 212, 286-287.

[252] Military pension records for Louis Anderson, Benjamin Bard and Douglas E. Lee (National Archives); Sayre, *Warriors of Color*, 16, 163; Schubert, *On the Trail of the Buffalo Soldier II*, 131.

[253] Military pension records for William Webb, William Brent, Jacob Watkins and Webb Chatmoun (National Archives); Sayre, *Warriors of Color*, 46-47, 49, 460-463; Schubert, *On the Trail of the Buffalo Soldier II*, 29, 36, 148.

[254] Military pension record for George Foster (National Archives); Sayre, *Warriors of Color*, 35-37, 43, 51, 54, 58-60.

[255] Robert W. Peterson, *Only the Ball Was White* (New Jersey: Prentice-Hall Inc., 1970), 251-252; Burlington *Free Press*, December 22, 1972; Vermont Death Records, 1909-2008, George Osborne.

[256] Unidentified manuscript in miscellaneous files, Troop H, Tenth Cavalry, RG 98 (National Archives).

BIBLIOGRAPHY

MANUSCRIPTS

Brooks, John B. Interview with Major General John B. Brooks, Fort Huachuca Museum files, Fort Huachuca, Arizona.

Fort Davis National Historic Site. Fort Davis, TX: Document file on "Fort Davis – Subposts," document file on "Lieutenant Levi P. Hunt," letter of First Lieutenant Robert G. Smither, Acting AAG, Tenth Cavalry to Captain Carpenter June 4, 1878, Report of Operations, Company H, Tenth Cavalry from May 20 to August 29, 1878, Captain L. H. Carpenter.

Grierson, Robert K. "Journal Kept on the Victorio Campaign in 1880" (typescript in Fort Concho Library and Archives).

"Proceedings of the Fifth Annual Reunion of the Survivors of the Sixth US Cavalry," July 3, 1888, Heinrich G. Mueller, Secretary.

Robinson, Warren. correspondence and telephone interview with Warren Robinson, great-grandson of Reuben Waller.

U.S. Army Continental Commands, 1821-1920. District of the Pecos Records, 1878-1881. Record Group 393, National Archives.

U.S. Army Continental Commands, 1821-1920. Letters Sent. Division of the Missouri, 1866-1875, Department of Texas, 1875-1885, Record Group 98, National Archives.

U.S. Army Continental Commands, 1821-1920. Muster rolls for Company H, Tenth U.S. Cavalry, August 1867 - December 1905. Record Group 98, National Archives.

U.S. Bureau of the Census. Ninth, Tenth, Eleventh, Twelfth, Thirteenth and Fourteenth Census of the United States, Population Schedules.

U.S. Bureau of Pensions. Military Pension Records for Henry Allen, Louis Anderson, Charles Black, William Black, Benjamin Bard, William Brent, Webb Chatmoun, James H. Clayton, Michael Finnegan, George Foster, George Garnett, George Goldsby, Washington Hardaway, William Hawkins, Dorsey Johnson, Douglas E. Lee, Colonel E. Miller, James H. Thomas, Jacob Watkins, Asa Weaver, William Webb, Robert White, Jacob Young.

U.S. Department of War, Adjutant General's Office. Letters Received, 1866-1916, Record Group 98, National Archives.

U.S. Department of War, Appointment, Commission and Promotion files, Charles G. Ayres, Charles T. Boyd, Louis H. Carpenter, Charles L. Cooper, James S. Greene, William R. Harmon, Levi P. Hunt, Eugene P. Jervey, Jr., Alexander S.B. Keyes, William Shipp. Record Group 94, National Archives.

U.S. Department of War, Carded Medical records for Alexander Adams, William C. Alexander, James Andrews, Benjamin Banks, Charles Barnum, John Billings, Charles Black, Joseph Claggett, George W. Forster, James Jackson, Dorsey Johnson, George Johnson, James Johnson, Silas Jones, John Lisby, Colonel Miller, John Morgan, John Muchs, Alfred Owings, Simon Peter, James H. Thomas, Henry Walker and Reuben Waller. Record Group 94, National Archives.

U.S. Department of War, Medal of Honor file for Dennis Bell, Tenth Cavalry, Record Group 94, National Archives.

U.S. Department of War, Records of the Judge Advocate General. Military Court Martial Records for Henry Allen, Louis Bell, William Booter, Alexander Brown , George Bumferts, William Chase, Ellson Clark, William Clark, Andy Clayton, James Clayton, Pollard Cole, Amos Cormack, John H. Curtiss, Alfred Dixon, George Douglass, Robert Ellis, William Epps, Michael Finnegan, Charles Gray, William Hamlin, Benjamin Harris, Lewis Hayes, John Henry, Dickson Hunter, Charles Jefferson, William Johnson, Silas Jones, Doc Mocfield, William Oliver, William Pierce, Samuel Porter, Charles Reed, Greenfer Shanklin, Ephraim Smith, Robert

Valentine, Anderson Wilson. Record Group 153, National Archives.

U.S. Department of War, Records of U.S. Army Commands. Registers of Enlistments in the United States Army, 1798-1914, Tenth Cavalry, Microcopy 233, Record Group 94, National Archives.

U.S. Department of War, Regimental Returns, Tenth Cavalry, 1866-1916, Microcopy 774, Record Group 94, National Archives.

U.S. Department of War, Returns of United States Military Posts, 1800-1916. Fort Riley, KS, January - February 1868; Camp Supply, IT, August 1869 - December 1870; Fort Sill, IT, May - October 1871, May 1873 - March 1875; Fort Gibson, IT, September - November 1872, Fort Davis, TX, April 1875 - April 1885. Microcopy 617, Record Group 94, National Archives.

BOOKS

Brady, Cyrus Townsend. *Indian Fights and Fighters.* Lincoln, NE: University of Nebraska Press, 1971.

Burton, Art T. *Black, Buckskin and Blue: African American Scouts and Soldiers on the Western Frontier.* Austin, TX: Eakin Press, 1999.

Carlson, Paul H. *The Buffalo Soldier Tragedy of 1877.* College Station, TX: Texas A & M University Press, 2003.

Carpenter, Edward and Louis Henry. *Samuel Carpenter and His Descendants.* Philadelphia: J.B. Lippincott Company, 1912.

Carriker, Robert. *Fort Supply, Indian Territory: Frontier Outpost on the Plains.* Norman, OK: University of Oklahoma Press, 1970.

Carter, Robert G. *On the Border with Mackenzie.* New York: Antiquarian Press, Ltd., 1961.

Cashin, Herschel V. *Under Fire with the Tenth U. S. Cavalry.* Niwot, CO: University Press of Colorado, 1993.

Chambers, Clint E. and Carlson, Paul H. *Comanche Jack Stilwell: Army Scout and Plainsman,* Norman: University of Oklahoma, 2019.

Federal Writers' Project: Slave Narratives, vol. 16 (Washington, DC, 1941), interview with Madison Bruin

Fisher, Barbara E. "Forrestine Cooper Hooker's Notes and Memoirs on Army Life in the West, 1871-1876." unpublished thesis: University of Arizona, 1963.

Fletcher, Marvin, *The Black Soldier and Officer in the United States Army, 1891–1917*(Columbia: University of Missouri Press, 1974).

Glass, E.L.N. *The History of the Tenth Cavalry, 1866-1921.* Fort Collins, CO: The Old Army Press, 1972.

Goodrich, S. G., *A Pictorial History of the United States* (New York: Huntington and Savage, Mason and Law, 1860).

Haley, James L. *The Buffalo War: The History of the Red River Uprising of 1874.* Garden City, NY: Doubleday and Company, 1976.

Harwood, Thomas. *History of New Mexico Spanish and English Missions of the Methodist Episcopal Church From 1850-1910*, Vol. 1. Albuquerque, NM: El Abogado Press, 1908.

Heitman, Francis B. *Historical Record and Dictionary of the United States Army From its Organization, September 29, 1789 to March 2, 1903.* Washington, DC: Government Printing Office, 1903.

Hooker, Forrestine Cooper. *When Geronimo Rode.* Garden City, NY: Doubleday, Page and Co., 1924.

Hooker, Forrestine Cooper. *Child of the Fighting Tenth: On the Frontier with the Buffalo Soldiers,* New York: Oxford University Press, 2003, ed. Steve Wilson.

Jackson, Jr., Jesse, *A Social History of the Tenth Cavalry, 1931-1941,* (Fort Leavenworth, KS: US Army Command and General Staff College, 1976).

Kinevan, Marcus E. *Frontier Cavalryman: Lieutenant John Bigelow with the Buffalo Soldiers in Texas.* El Paso, TX: Texas Western Press, 1998.

Lanning, Michael Lee. *The African American Soldier: From Crispus Attucks to Colin Powell.* Secaucus, NJ, 1997.

Leckie, William H. *The Buffalo Soldiers: A Narrative of the Black Cavalry in the West.* Norman, OK: University of Oklahoma Press, revised 2003.

Leckie, William H. and Shirley A. *Unlikely Warriors: General Benjamin H. Grierson and His Family.* Norman, OK: University of Oklahoma Press, 1984.

Merrill, Samuel, *The Seventieth Indiana Volunteer Infantry in the War of the Rebellion* (Indianapolis: The Bowen-Merrill Company, 1900).

Minton, John, *The Houston Riot and Courts-Martial of 1917* (Munguia Printers: San Antonio, TX, undated).

Murray, Arthur, *A Manual for Courts-Martial,* (New York: John Wiley and Sons, 1893).

Official Army Register for 1900. Washington, DC: Adjutant General's Office, 1900.

Peterson, Robert W. *Only the Ball Was White*. New Jersey: Prentice-Hall Inc., 1970.

Powell, William H. *Records of Living Officers of the United States Army*. Philadelphia, 1890.

Quackenboe, G. B., *Illustrated School History of the United States* (New York: D. Appleton and Company, 1868).

Record of Engagements with Hostile Indians within the Military Division of the Missouri from 1868 to 1882. Washington: Government Printing Office, 1882.

Remington, Frederic. *Selected Writings*, compiled by Frank Oppel. Secaucus, NJ: Castle, 1981.

Rodenbough, Theophilus F and Haskin, William L., *The Army of the United States: Historical Sketches of Staff and Line with Portraits of Generals-in-Chief*. New York: Maynard, Merrill and Company, 1896.

Roster of Non-commissioned Officers of the Tenth U.S. Cavalry. Bryan, TX and Mattituck, NY: J.M. Carroll and Company, 1897.

Sayre, Harold Ray. *Warriors of Color*. Fort Davis, TX, 1975.

Schubert, Frank N. *Black Valor, Buffalo Soldiers and the Medal of Honor, 1870-1898*. Wilmington, DE: Scholarly Resources, Inc, 1997.

Schubert, Frank N. *On the Trail of the Buffalo Soldier: African Americans in the US Army*. Wilmington, DE: Scholarly Resources Inc, 1995.

Schubert, Irene and Frank N. *On the Trail of the Buffalo Soldier II: New and Revised Biographies of African Americans in the U.S. Army, 1866-1917*. Lanham, MD, Toronto, Oxford: The Scarecrow Press, 2004.

Smith, Cornelius C., *Fort Huachuca:* The *History of a Frontier Post* (Fort Huachuca Museum, 1977).

Sonnichsen, C.L. *The El Paso Salt War.* El Paso, TX: Texas Western Press, 1961.

Steward, Theophilus G. *The Colored Regulars in the United States Army.* New York: Arno Press and the New York Times, 1969.

The Battle of Beecher Island, Fought - September 17, 18, 1868. Wray, CO: Beecher Island Battle Memorial Association, revised printing, Sterling, CO: Royal Printing Co, 1985.

The War of the Rebellion: A Compilation of the Official Records of the Union and Confederate Armies, Series 1, v 45 parts 1 and 2. Washington, DC: U.S. Government Printing Office, 1897

Thrapp, Dan L. *Victorio and the Mimbres Apaches.* Norman and London: University of Oklahoma Press, 1974.

Wallace, Ernest, ed., *Mackenzie's Official Correspondence, 1873-1879* (Lubbock, TX: West Texas Museum Association, 1967).

Wharfield, Colonel Harold B. *With Scouts and Cavalry at Fort Apache.* ed John Alexander Carroll, (Tucson, AZ: Arizona Pioneers' Historical Society, 1965).

Wharfield, Colonel Harold B. *10th Cavalry and Border Fight.* (El Cajon, CA, 1965).

Wittenberg, Eric J. *Gettysburg's Forgotten Cavalry Action.* (Gettysburg, PA: Thomas Publications, 1998).

Wittenberg, Eric J. *The Battle of Brandy Station: North America's Largest Cavalry Battle*, (Charleston, SC: The History Press, 2010).

Wooster, Robert, *Soldiers, Sutlers and Settlers: Garrison Life on the Texas Frontier* (College Station: Texas A & M University Press, 1987).

Work, David K., "The Fighting Tenth Cavalry: Black Soldiers in the United States Army 1892-1918," unpublished thesis: Oklahoma State University, 1998.

ARTICLES

Buechler, John, *"Buffalo Soldiers in the Green Mountains," Chittenden County Historical Society Bulletin*, 5:2, 3 (November 1969 and April 1970).

Finley, James P., "The Buffalo Soldiers at Fort Huachuca," *Fort Huachuca Illustrated*, Volume 2, (1996).

King, Joseph H. "Brief Account of the Sufferings of a Detachment of United States Cavalry from Deprivation of Water." *The American Journal of the Medical Sciences.* 75 (1878).

Mooar, J. Wright. "Frontier Experiences of J. Wright Mooar," *West Texas Historical Association Year Book.* Vol. IV. (June 1928).

Rister, Carl C. "Early Accounts of Indian Depredations," *West Texas Historical Association Year Book.* Vol. II. (June 1926).

Schubert, Frank N., "Black Soldiers on the White Frontier: Some Factors influencing Race Relation," *Phylon* 32 (1971).

Waller, John L., "Colonel George Wythe Baylor," *Southwestern Social Science Quarterly* 29 (September 1950).

Work, David K. "The Buffalo Soldiers in Vermont, 1909 – 1913," *Vermont History,* 73 (Winter/Spring 2005).

Work, David K. "Enforcing Neutrality: The Tenth U.S. Cavalry on the Mexican Border, 1913-1919," *Western Historical Quarterly*, Vol. 40, No. 2 (Summer, 2009).

NEWSPAPERS AND PERIODICALS

Army-Navy Journal, April 7, 1866, April 28, 1866, May 4, 1867, September 7, 1867, April 11, 1868, February 6, 1869, March 24, 1869, February 26, 1870, January 19, 1884.

Barre Daily Times, October 11, 1911 and February 29, 1912.

Bennington Evening Banner, October 11, 1911

Bisbee Daily Review, July 4, 1919.

Burlington *Free Press*, December 22, 1972.

Harper's Weekly, September 7, 1867.

Hutchinson *News Herald*, August 5, 1940.

Illinois *Record*, December 3, 1898.

Lincoln *Sentinel-Republican*, June 30, 1932.

New York *Times*, November 28, 1880.

Omaha *Bee*, April 1 and September 18, 1883.

Richmond *Planet*, January 29, February, 19, May 14, July 16, July 23, November 19, November 26, December 13 and December 17, 1898.

San Antonio *Express*, August 22, 1877, March 28, 1916.

San Antonio *Light*, April 16, April 19 and May 3, 1898.

The Norwalk Hour, Oct. 11, 1911.

The Arizona Republican, July 4, 1919.

Winners of the West. Vol. 1: July 1924, October 1924. Vol. II:
August 1925. Vol. IV: March 1927. Vol. V: August 1928. Vol.
VI: May 1929. Vol. VII: March 1930, November 1930. Vol. XI:
February 1934, April 1934. Vol. XIV: July 1937.

APPENDIX

Roster of Company H Members who Enlisted in 1867

Alexander Adams   1867-69 Born in Maryland in 1845. (5'5")
Served in Fifth Massachusetts Cavalry, enlisted in Tenth Cavalry at
Washington, DC in spring 1867. Medical discharge April 7, 1869
for heart disease. He died on post April 9, 1869 of acute
rheumatism.

Henry Allen   1867-97 A mulatto born 1849 in Richmond,
Virginia, he enlisted in Company F, 20th USC Infantry at the age of
sixteen. After he mustered out in New Orleans in October 1865, he
tried farming for a time and then joined the Tenth Cavalry. Served
as company gardener, promoted to Sergeant. Court-martialed
twice. Despite continuing attitude problems, Allen served for thirty
years in Troop H, retiring at Fort Assiniboine on October 31, 1897.
He remained in Havre, Montana farming and applied for a pension
in 1913. Married Martha in 1888, had daughters Lizzie and Isabel,
son Charles.

+John Allen   1867 Born in 1845 at Memphis, Tennessee.(5'10")
Listing his occupation as engineer, he enlisted at Memphis in 1867.
Listed sick in September, he received a disability discharge at Fort
Riley on December 15, 1867.

+Robert B. Banks   1867-88   Banks, a waiter from Halifax,
Pennsylvania, enlisted with Captain Carpenter on June 20, 1867 at
Harrisburg, PA.(5'10".) He would serve with the regiment for
twenty years as a farrier and sergeant.   On December 15, 1888,
Banks died of exhaustion, gastritis and old age.

+Archer Bates   1867 Born 1846 in Lynchburg, Virginia, farmer.
(5'2") Discharged December 25, 1867 for disability at Fort Riley.

Isaac A. Beckwood 1867-72 Joined December 1867 to replace
troopers who had received disability discharges. Born 1844 in
Johnson County, North Carolina, farmer. (5'6")

+John Billings   1867-72 Born 1846 in New York, servant. (5'5")
Enlisted at New Haven, Connecticut. Suffered from frostbite in
1868, served as company cook.

+Charles Black   1867-71 Black was born at Richmond in 1841. In
September 1863 he joined 54th Massachusetts Colored Volunteers.
Tried for sleeping on guard duty in September 1864 and sentenced
to confinement at Fort Marion, Florida. Black joined the Tenth
Cavalry on June 11, 1867 and served as Company H cook. Black
suffered from scurvy during the winter campaign of 1868. At Fort
Sill in 1870, Black's vision in his right eye was destroyed when
Bartlett Mutes accidentally discharged a shotgun that he believed to
be unloaded. Black received a disability discharge on March 28,
1871. Black continued to have problems with his right eye. When
his left eye developed a cataract he could "barely distinguish light"
and had to be led around by an attendant. The Bureau of Pensions
did reconsider its refusal in Black's case and recommended that his
injury be considered in the line of duty and his claim allowed since
at the time of the injury "he was in his proper place and doing
nothing that he should not do."

+William Blaids   1867-68 Born in Kentucky, 1842, laborer, (5'8")
Enlisted at Louisville, deserted on May 26, 1868.

+Frank Bloodson   1867 Born 1845 in Atlanta, Georgia, farmer.
(5'7") Enlisted at Memphis, sick in hospital September 1867,
disability discharge at Fort Riley December 15, 1867.

+William Bradshaw   1867-

+John Brown   1867-72 Born in Montgomery County, Tennessee
in 1843, waiter. (5'5") Discharged June 26, 1872 at Fort Sill.

+George Bumpferts   1867-72 Born in Fayette County., Tennessee
in 1846, laborer. (5'3") Enlisted at Memphis in June 1867.
Sentenced for being found asleep on stable guard. Discharged on
June 27, 1872 at Fort Sill.

+Charles Burns   1867-76 Promoted to sergeant during second
enlistment.

+Henry Carpenter 1867-68 Born 1832 in Philadelphia, Pennsylvania and enlisted there in June 1867, barber. (5'5") Disability discharge on July 29, 1868 at Fort Riley.

+John Carpenter 1867-72 Born in 1846 in Clinton County, Missouri, farmer. (5'9") enlisted at Fort Leavenworth in May 1867.

+John M. Claggett 1867-73 Born in 1845 in Prince George County, Maryland, farmer. (5'5") Enlisted at Washington, DC on July 9, 1867. Wounded in the leg with an arrow at Beaver Creek. Suffered frostbite in the blizzard of 1868. Deserted on July 26, 1873 while under suspicion of petty larceny. Claggett was apprehended at Fort Gibson on October 7. He pled guilty, received a five year penitentiary sentence and a dishonorable discharge.

+Joseph Claggett 1867-97 Born in 1846 in Prince George County, Maryland, laborer. (5'5") Enlisted at Washington, DC on July 9, 1867. Suffered frostbite in the blizzard of 1868. Served thirty years as a saddler and sergeant. Retired on September 20, 1897 and settled in the town of Havre, Montana just outside of Fort Assiniboine.

+James Clayton 1867-72 Born in 1844 in Richmond, Virginia, shoemaker. (5'5") Had served as corporal in the 84th USC Infantry during the Civil War, enlisted at Washington, DC in July 1867. Promoted to sergeant. At Camp Supply in 1870, while sergeant of the guard, he allowed Private James Henry of Company A to escape from the guardhouse. Forfeited $5 per month for three months.

+Pollard Cole 1867-94 Born in 1842 in Georgetown, Kentucky. Enlisted on June 14, 1867 at Louisville, Kentucky citing prior service in the 12th USC Heavy Artillery, Company K. (5'6") Served as a farrier and promoted to sergeant in 1872. Cited for gallantry in action at Fort Sill 1874. Married while at Fort Davis. Found sleeping on his post at Fort Davis in 1879, he was sentenced to be confined in the guardhouse for two months and forfeit $10 per month. Member of the patrol that captured Mangus. He retired on August 13, 1894 after twenty-seven years of service. Moved to El Paso, Texas with his wife, the former Estephana Gonzales. He died on May 20, 1900 while away from home visiting family in Georgetown, Kentucky. Estephana spent the next twelve years trying to secure a government pension to benefit their son Joseph Pollard.

+Amos Cormack   1867- 69   Born in 1845 in Gettysburg, Pennsylvania and enlisted there, carpenter. (5'5") Promoted to First Sergeant in 1868. Deserted on August 6, 1869, with sixteen of his men. A thirty dollar bounty was offered for the capture of each man, but Cormack was never found.

+Perry Curry 1867 Born in 1846 at Selma, Alabama, laborer. (5'6") Enlisted at Memphis, disability discharge on December 25, 1867 at Fort Riley.

+Thomas Daniels   1867-72 Born in 1846 in Bedford County, Tennessee, laborer. (5'6") Enlisted at Louisville, Kentucky.

+Dunn Day 1867   Born in 1846 at Jackson, Tennessee, farmer. (5'4") Enlisted at Memphis, disability discharge at Fort Riley on December 15, 1867.

+Alfred Dixon   1867-70 Born in 1845 in Georgia. (5'4") Enlisted at Memphis, promoted to corporal in 1868. On August 6, 1869, he deserted from Camp Supply with sixteen other soldiers.   Dixon finally surrendered himself in October. He received a reduction in rank, a dishonorable discharge and four years confinement at Jefferson City, Missouri on January 10, 1870..

+Robert Edwards   1867-70   Born in 1846 at Louisville, Kentucky, cook. (5'7") Enlisted at Memphis, deserted April 24, 1870, captured June 4, dishonorable discharge on November 20.

+Thomas Edwards   1867-72 Born in 1846 in Kentucky, laborer. (5'8") Enlisted at Louisville, discharged Fort Sill.

+Charles Evans 1867-72 Born in 1846 in Alabama, farmer. (5'5") Enlisted at Memphis.

+Frank Fields   1867-72 born in 1845 at Anderson, South Carolina, laborer. (5'4") Enlisted at Memphis.

+Ezariah S. Freeman   1867-68 Born in 1843 at Sheffield, Massachusetts, shoemaker. (5'5") Former infantryman in the 29th Connecticut Infantry, Colored, Co D. Enlisted in Connecticut, deserted March 16, 1868.

James W. Gibson   1867-87 Born in 1846 at St. Charles, Missouri, laborer. (5'5") Member of the patrol that captured Mangus.

+Richard Gowans   1867-69 Born in 1843 in Philadelphia, Pennsylvania, laborer. (5'4") Enlisted at Harrisburg, deserted August 6, 1869.

+Daniel Grissom   1867-68 Born in 1845 in Wilson County, Tennessee, farmer. (5'6") Enlisted at Memphis, disability discharge on April 11, 1868

+Henry Grose   1867-68 Born in 1843 in Missouri, laborer. (5'3") Enlisted at Washington, DC, dishonorable discharge, confinement for life Missouri state prison.

+John Homager   1867 Born in 1837 at Pottstown, Pennsylvania, laborer. (5'10") Enlisted at Philadelphia by Captain Carpenter as a sergeant because he could read and write. Original First Sergeant of Company H. Sick in hospital at Fort Harker in September 1867, disability discharge on December 25, 1867.

+Henry Harper   1867 Born in 1846 at Cherokee, Alabama, farmer. (5'7") Enlisted at Memphis, had prior service with Captain Carpenter as a first sergeant in the Fifth USC Cavalry. Died of an unknown disease on October 27, 1867 at Big Creek, Kansas.

+George Harris   1867-68 Born in Georgia in 1844, farmer. (5'10") Enlisted at Memphis, disability discharge on April 11, 1868.

+Thomas Hayden   1867-68 Born in 1844 at Brunswick, Maine, laborer. (5'6") Enlisted at Portland, Maine, veteran of the 54th USC Infantry. Promoted to corporal, deserted on February 29, 1868.

+John Hensley   1867-

+Jerry Hogins   1867   Born in 1840 at Spartanburg, South Carolina, laborer. (5'5") Enlisted at Louisville, died of cholera on September 3, 1867.

+George Holtzinger   1867-69   Born at Safe Harbor, Pennsylvania in 1846, farmer. (5'10") Enlisted at Harrisburg, Pennsylvania on June 20, 1867, disability discharge on March 27, 1869 at Fort Wallace. His widow Hannah filed for a pension on July 24, 1869.

+Samuel Jackson   1867- . Enlisted at Memphis, Tennessee in June 1867.

+Ernie Johnson   1867-   In confinement at Fort Harker, dishonorable discharge September 1867.

+William Johnson   1867-69   Deserted on August 6, 1869, captured on August 9, confined at hard labor, dishonorable discharge.

Alfred Jones   1867-77   Born in 1845 in Pennsylvania, recruited by Captain Carpenter in Philadelphia, farmer. (5'5")

+Benjamin Jones   1867-   Sick in hospital in September 1867.

Joseph Jones   1867-69   Born in 1844 in Franklin County, Pennsylvania, farmer. (5'7") Died on September 10, 1869 of typhoid fever at Fort Sill.

+Manuel Jones   1867   Born at Shelby, Kentucky in 1846. He became the first Company H deserter when he slipped away from a patrol on August 13.

+Mitchell Jones   1867   Born in 1845 at Atlanta, Georgia, farmer. (5'4"). Sick in hospital at Fort Harker, Kansas on September 23, 1867, died September 24 of severe inflammation of the jaw and peritonitis.

Philip Jones   1867-97   Born in 1846 at Fredericksburg, Virginia, a servant. (5'7") Retired as a corporal after thirty years of service on November 13, 1897 at Fort Assiniboine. His widow Mollie Jones filed for a pension on March 14, 1912 in Montana.

+Silas Jones   1867-97   Born in Saline County, Missouri in 1846. (5'5") Enlisted in Memphis, a veteran of the Sixth USC Cavalry. Served as trumpeter. Had gonorrhea in 1877 and syphilis in 1880. On April 23, 1870 at Camp Supply, he fell asleep on duty while guarding the hay stacks. He was released without punishment. Present at the battle of Rattlesnake Springs and the capture of Mangus. Retired after thirty years of service on July 7, 1897. Married and lived with his wife Nellie in Havre, Montana.

Robert Kurney   1867-98   Born at Richmond, Virginia in 1846, farmer. (5'7") Served as a farrier and sergeant. Retired on June 29, 1898 after thirty-one years of service. Died at 6:20pm September 1, 1911 at Walter Reed General Hospital, Washington, DC, buried at Arlington National Cemetery.

+Charles Lewis   1867-72   Born in 1846 at Nashville, Tennessee, laborer. (5'5") Enlisted at Memphis.

+George Lewis   1867-82   Born in 1845 in Virginia, laborer. (5'10") Veteran of the 44th USC Infantry, later served with Company L, Tenth Cavalry.

+George Martin   1867-69   Born in 1842 at Richmond, Virginia, farmer. (5'5") Enlisted at Memphis, disability discharge on May/26, 1869 at Fort Wallace.

+Ephraim McCutchen   1867-72   Born in 1846 at Nashville, Tennessee, laborer. (5'9")

+Allen McPherson   1867-69   Born in 1845 at Martinsburg, Virginia, porter. (5'6") Enlisted at Detroit, deserted on August 6, 1869. Former corporal in the 64th USC Infantry.

+David H. Mead   1867-69   Born in 1846 at Lancaster, Pennsylvania, waiter. (5'5") Enlisted at Harrisburg, Pennsylvania, deserted on August 6, 1869.

+Peter E. Moore   1867-68   Born in 1844 at Canton, Ohio, laborer. (5'8") Enlisted at Detroit, disability discharge on July 31, 1868 near Fort Wallace.

+John Moore   1867-72   Born in 1846 in Indiana, farmer. (5'8")

+Joseph Murphy   1867-75   Born in 1846 at Alexandria, Virginia, painter. (5'10") Re-enlisted in Co K, dishonorable discharge on September 24, 1875

+Bartlett Mutes   1867-72   Born in 1845 in Meyer County, Kentucky, laborer. (5'9") Enlisted at Louisville, discharged in 1872 at Fort Sill.

+William Oliver   1867-69   Deserted from Camp Supply on August 6, 1869, captured within a matter of days, received a prison sentence and a dishonorable discharge.

+Simon Peter   1867-72   Born in 1846 in Georgia, farmer. (5'6") Enlisted at Memphis. Married Louisa Jones in January 1891.

+William Pierce   1867-72   Served as a trumpeter. Stole a cap valued at $4 from the sutler's store at Camp Supply in 1870. Sentenced to confinement at hard labor under guard for six months and forfeiture of $12 per month for same period.

+Scott Pinkley   1867-69   Born in 1845 at Charleston, South Carolina, laborer. (5'5") Enlisted at Memphis, disability discharge on May 26, 1869.

+John D. Price   1867-76   Born at Harrisburg, Pennsylvania in 1846, enlisted there on July 3, 1867, shoemaker. (5'6") Dishonorable discharge on January 2, 1876.

Toney Ratcliff   1867-82   Born in Missouri in 1846, enlisted on May 25, 1867 at Fort Leavenworth, saddler. Veteran of the 61st USC Infantry, promoted to sergeant in Company H.

+Frank Rogers   1867-68   Born in 1846 at McMillen, Tennessee, tailor. (5'4") Enlisted at Memphis, deserted May 26, 1868.

+John Rose   1867-69   Born in 1846 in Grant County, Kentucky, laborer. (5'8") Enlisted at Detroit, sick in hospital September 1867, deserted August 1, 1869.

William Ross   1867-70   Born in 1841 in Philadelphia, Pennsylvania, tailor. (5'7") Deserted on March 16, 1870, apprehended on June 2.   Received a dishonorable discharge on November 24, 1870 and three years confinement.

+John Royster   1867-72   Born in 1846 in Henry County, Missouri, laborer. (5'10") Enlisted at Fort Leavenworth.

+Charles Sampson (Sawyers)   1867-72   Born in 1846 at Knoxville, Tennessee, laborer. (5'4") Enlisted at Memphis.

+Sidney Sanders   1867-68   Born in 1840 in Owen County, Kentucky, baker. (5'11") Enlisted at Washington, DC. Deserted along with Ezariah Freeman and James Wright on February 29, 1868, captured March 24, 1868.

+Charles Shavers 1867-69   Born in 1845 at Lynchburg, Virginia, laborer. (5'6") Enlisted at Memphis, deserted August 6, 1869, captured August 12, 1869, deserted August 17, 1869, captured September 2, 1869, deserted September 12, 1869.

Shelvin Shropshire   1867-1906   Born in Alabama in 1847. Veteran of the 15th USC Infantry, enlisted in Company C in 1867. Promoted to First Sergeant of Troop H in 1891.   Died on February 7, 1906 at Fort Mackenzie, Wyoming, reinterred at Custer Battlefield, Montana.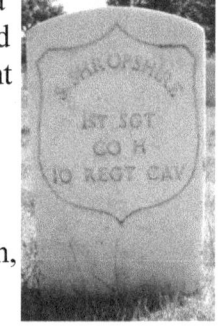

+Amos Smith   1867-68 Born in 1845 at Charleston, South Carolina, farmer. (5'5") Enlisted at Memphis, deserted on January 14, 1868.

+John Stevens   1867-72   Born in 1846 at Cherokee, Georgia, laborer. (5'10") Enlisted at Memphis, promoted to First Sergeant.

+Wood Solon Taylor   1867-72   Born at New Haven, Kentucky in 1845, laborer. (5'5") Enlisted at Louisville.

+James J. Thistle   1867-69   Born in 1846 at Milford, Delaware, sailor. (5'4") Enlisted at New Haven, deserted on August 6, 1869.

+James H. Thomas 1867-97 Born in 1846 at James City, Virginia, barber. (5'6") Trumpeter, had served as a musician in the First USC Infantry. Suffered from scurvy during the winter campaign of 1868. Appointed Chief Trumpeter of the Tenth Cavalry in 1891 and Chief Musician on July 7, 1897. Retired after thirty years of service on August 3, 1897.

+John Thompson 1867-69 Born in 1846 at Memphis, Tennessee, farmer. Enlisted at Memphis, served as a saddler. Deserted on August 11, 1868, captured on March 9 1869, deserted on March 29, 1869 captured on April 1, 1869, dishonorable discharge on April 15, 1869.

+Robert Thrash 1867-70 Born in 1845 in Houston County, Georgia, farmer. (5'4") Enlisted at Memphis, disability discharge in April 1870. Filed for pension on June 22, 1882, widow Adeline Branham filed for pension on March 3, 1898 in Georgia.

+Butler Tillman 1867-69 Born in 1846 in Tennessee, farmer. (5'10") Enlisted at Memphis, deserted on January 18, 1868 with Amos Smith. Captured on April 28, 1868, processed at Fort Hayes on May 4, 1868, deserted on September 12, 1868, captured on September 16, 1869, dishonorable discharge on October 16, 1869 at Camp Supply.

+Reuben Waller 1867-72 Born on January 5, 1840 in Platte County, Kentucky, farmer and distiller. (5'8") Enlisted at Fort Leavenworth. Served with the Confederate cavalry in 29 battles. Suffered an accidental gunshot wound on July 28, 1871. Mustered out in 1872 and worked as a civilian employee for the army. Applied for and received a pension on October 28, 1875. Died August 20, 1945 in El Dorado, Kansas at the age of 105.

+George D. Washington 1867-68 Born in 1845 in Buckingham County, Virginia, laborer. (5'5") Enlisted at New York, disability discharge on July 31, 1868.

+James Wilkerson   1867-

+Charles Williams   1867   Born in 1846 in North Carolina, laborer. (5'6") Deserted on December 14, 1867.

+Daniel Williams   1867-72   Born in 1846 at Harrisburg, Pennsylvania, blacksmith. (5'7") Enlisted at Philadelphia.

+Jerry Williams   1867-68   Born in 1846 in Hardman County, Tennessee, laborer. (5'5") Enlisted at Memphis, convicted of murder on March 6, 1868.

+Joseph W. Williams   1867-68   Born in 1846 at Memphis, Tennessee, laborer. (5'3") Received a disability discharge on June 29, 1868.

+Osborne Williams   1867-72   Born in 1848 in Berks County, Georgia, servant. (5'8") Enlisted at Memphis, promoted to corporal.

+James Wilson   1867-72   Born in 1846 in Mississippi, farmer. (5'6") Enlisted at Memphis.

+James Wright   1867-69   Born in 1846 at Richmond, Virginia, farmer. (5'9") Enlisted at Detroit, served on detached duty as a teamster. Deserted along with Ezariah Freeman and James Wright on February 29, 1868, captured March 24, 1868, deserted August 6, 1869.

+Jacob Young   1867-80   Born in 1846 at Elizabethtown, Kentucky, laborer. (5'8") Enlisted in Louisville, Kentucky on May 24, 1867. Veteran of the 24th USC Infantry, promoted to corporal in 1872, First Sergeant of Company H. Retired and settled near Fort Gibson with his wife Polly. Suffered from rheumatism and lumbago to such an extent that he could not maintain a job. Received a pension and remained at Fort Gibson until his death on September 22, 1916.

+ Indicates soldiers listed on the original July 31, 1867 muster roll.

Partial List of Company H Members who Enlisted after 1867

King Adams   1883-86   Born at Richmond, Virginia in 1835. Enlisted at San Antonio in December 1883, dishonorable discharge in February 1886.

James H. Alexander   1887- 1902   Promoted to sergeant in January 1898.

William C. Alexander   1872-74   Born in 1847 in New York City, waiter. (5'8") Treated for bronchitis, colic, fever, catarrh and rheumatism on a recurring basis. He finally received a certificate of disability and was discharged on August 22, 1874.

Charles S. Allen   1897-1900   Born in Virginia in 1866, laborer. (5'6") Discharged on July 26, 1900 at Fort Clark, joined 24th Infantry, discharged 1903. Married Clara Allen, Entered the Dayton, Ohio home for disabled volunteer soldiers with tuberculosis in 1910, died 1919.

William Allen   1882-85   Born in Tennessee in 1844, shoemaker. Veteran of the 12th USC Heavy Artillery. Had a revolver fall from his pocket, strike a table and discharge, wounding him in the right side while off duty at Fort Davis. Suffered from a serious inflammation of the nose and throat, also suffering from some deafness. He blamed exposure to the cold night air. Received a disability discharge in 1885. He married Fannie Clouds on April 21, 1902 at Lexington, Kentucky. Served as pastor of Union Baptist Church in Lebanon, Ohio and Zion Baptist in Paris, Kentucky. Applied for a pension but only received a minimal amount for deafness. Allen's condition continued to deteriorate until blood poisoning from his liver affected his mind. He was committed to the Eastern Kentucky asylum for the insane in November 1909 where he died March 10, 1910.

Clinton Anderson   1876- 80   Born in 1850 in Atlanta, Georgia porter in Covington, GA in 1870 waiter married, smoker 5'5" discharged 5/15/80 without character, theft, sentenced to 3 yrs at Huntsville, settled near Huntsville, wife Emily who had two children, four together, Iron worker

Creed Anderson 1877 Veteran of the 125th USC Infantry. Jennie Ingram, his widow, filed for pension in 1892 in California.

Louis Anderson 1876-81 Born a slave in Fort Valley, Georgia, worked as a teamster following his enlistment. He received a pension of $50 per month. Married in 1885 in Pecos County, Texas. Anderson died in El Paso, Texas on March 23, 1928 at the age of 81.

Robert Anderson 1898 Died of yellow fever at Siboney in August 1898.

James Andrews 1879-80 Suffered a serious concussion in December 1880 when another soldier beat him on the head with a club.

Aaron Archer 1873-75 Farmer, born in New Jersey. Enlisted at Philadelphia in January 1873. (5'8") Suffered frostbitten toes during the Red River War. Received a disability discharge on November 26, 1875.

Walter Armstrong 1885-86 Member of the patrol that captured Mangus.

Nathan Ashberry 1873 Deserted in October 1873 at Ft Sill.

Benjamin Banks 1882-84 Died of acute dysentery on November 25, 1884 at Fort Davis.

Benjamin Bard 1873-76 Born in 1853 in Pennsylvania. During the Red River War, he had frostbite so severe that he was allowed to ride in a supply wagon. He spent most of the remainder of his enlistment in the hospital and received a discharge on February 11, 1876 with a fifty percent disability. Married three times and had two sons and a daughter. He worked as a car sweeper for the Pennsylvania Railroad and the Baltimore & Ohio Railroad. Bard died at age 74 in Crestmont, Pennsylvania on August 19, 1927.

James S. Barnes   1896-1913 Born in 1874 in Maryland, laborer. (6') Discharged at Ft Ethan Allen September 20, 1913.

Charles Barnum   1868-73 Injured by falling from his horse on drill.

William P. Battle   1883-94 Member of the patrol that captured Mangus, promoted to corporal. Married Ellen Anderson and settled in Pittsburgh, Pennsylvania. He served in the militia with Battery B, Pennsylvania Light Artillery. Re-enlisted in the Tenth in 1898 and claimed to be the first bugler to sound the charge at San Juan Hill. When the body of President William McKinley passed through Pittsburgh following his assassination, Battle was chosen to blow taps. Worked for the Bureau of Mines of Pennsylvania. He was an officer of the Knights of Pythias, a member of the Veterans of Foreign Wars, and an active member of the Euclid Avenue AME Church. Battle died January 17, 1917 at his home in Pittsburgh from acute lobar pneumonia.

Dennis Bell   1897-1900 Enlisted with his brother Arthur in 1896. Awarded the Medal of Honor for the rescue at Tayabacao, presented on June 30, 1899. Promoted to corporal. Suffered from malaria while serving at Manzanillo. Lived and worked in the Washington, DC area after discharge. Bell died on September 25, 1953 and was buried at Arlington National Cemetery.

Louis Bell   1887-90 Blacksmith. In 1889, a woman named Frances Jones was killed in his tent at the San Carlos reservation by an accidental gunshot. Bell was acquitted. Deserted from Fort Apache on June 16, 1890. He was captured five miles from post the following morning, sentenced to confinement under guard at hard labor for two years and a dishonorable discharge.

Scott A. Berkley   1875-77 Born at Breckinridge, Kentucky in 1851, laborer. (5'7") Enlisted in April 1875, deserted February 16, 1877 captured two days later. Dishonorably discharged on April 30, 1877 at Fort Davis.

Randall Blunt   1882-87 Born in 1859 in Greenville, North Carolina, laborer. (5'6") Post gardener at Fort Davis. Developed rheumatism on patrol in Arizona. After one enlistment, he moved to San Francisco, married, and had twins, Randall K. and Thema K. He died April 5, 1922 at the home of his daughter Thema Simpson and was buried at San Francisco National Cemetery.

William Bly   1876-79   Born in 1855 at Gallatin, Tennessee, laborer. Received a disability discharge on October 7, 1879 at Fort Davis.

Solomon Boller   1882-92   Born in 1861 in Nottaway County, Virginia, blacksmith. (5'6") Company blacksmith. In 1885 in Bonita Canyon, Arizona, contracted catarrh of the stomach. Member of the patrol that captured Mangus. Married Fannie Thompkins in Gallup, New Mexico. During the 1890's, they lived in Albuquerque. Joseph Cammel, another Troop H soldier present at the capture of Mangus lived with the Bollers until his death on December 15, 1899. Received a pension in 1927. On November 12, 1951, his wife Fannie died. One month later, on December 23, 1951, Solomon Boller passed away in Los Angeles, California at age 90.

William Booter   1868-69   Born in 1846 at Donaldsonville, Virginia, laborer. (5'7") Ran a business on the side selling cavalry boots. Sentenced "to have one side of his head close shaven; to be dishonorably discharged and drummed out of the service." Received his dishonorable discharge on March 1, 1869 at Fort Wallace.

George Brasheal   1876-79   Born in 1853 at Williamston, South Carolina, laborer. (5'6") Received a disability discharge for a knife wound on May 5, 1879 at Fort Davis.

Edward H. Braxton   1895-99   Born in 1874 in New Kent, Virginia, laborer. (5'8") Promoted to sergeant. Discharged at Fort Sam Houston in 1899.

William Brent   1879-83   Born in Virginia on March 22, 1845, farmer. Veteran of the Fifth Massachusetts Colored Cavalry. Member of the patrol that located Victorio in August 1880. Moved to Leavenworth, Kansas in the 1890's and married Frances Lewis, had one son.

John R. Brooks 1892-98 Born in 1874 in Richmond, Virginia, printer. (5'5") Served as company clerk. Left camp to go into town without permission at Huntsville, Alabama on November 11, 1898. Shot and killed by local civilian, "Horse" Douglass who claimed to be in the pay of a conspiracy of white citizens intent on murdering members of the Tenth Cavalry as long as they remained in Huntsville.

Alexander Brown 1873-74 Born in 1851 in Washington County, Kentucky, laborer. (5'9") Deserted and stole weapons from the ordnance room to aid in his escape. Received a dishonorable discharge on July 19, 1874 at Fort Sill and confinement for five years.

David T. Brown 1888-98 Born in 1867 at Brooklyn, New York, waiter. (5'6") Promoted to corporal and sergeant. In 1898, he was working in the recruiting office in Atlanta.

Jerry M. Brown 1876-81 Born in LaGrange, Georgia in 1855, waiter. (5'9") Promoted to corporal. Discharged at Pena Colorado on October 27, 1881. Married Arrelia, hired out as a farm worker in Georgia and Alabama.

Thomas Bruff 1885-98 Born in 1862 at Morristown, Pennsylvania, hostler. (5'7") Member of the patrol that captured Mangus. Went to Cuba with the regimental band, but did not return. He became one of the many victims of yellow fever.

Nathan Bullock 1896-99 Born in 1874 in Stovall, North Carolina, farmer. (5'9") Engaged with selected members of the Third Squadron to reinforce and resupply Cuban insurgents.

Henry Burns 1869-72 Born at Guilford, North Carolina in 1848, laborer. (5'8") Deserted on July 31, 1871.

Scott Cain 1883-85 Born in 1862 in Brookhaven, Mississippi, laborer. (5'5") In January 1885, Scott Cain, who had sold his greatcoat, entered the barracks during a formation and walked out with Peter Dehoney's coat. Court-martialed for the theft and for being absent without permission. He received a dishonorable discharge and confinement at hard labor for one year.

Joseph Cammel   1879-89   Born in Madison County, Kentucky in 1856, laborer. (5'7") Veteran of the 2nd USC Light Artillery. Member of the patrol that captured Mangus. Discharged at Fort Apache. Lived in New Mexico with Solomon Boller when pension filed.

James Campbell   1867-77   Born in Richmond County, North Carolina in 1847, farmer. (5'4") Served his first enlistment in Company D, promoted to sergeant. Led a detachment trailing Apaches through the Carrizo and Guadalupe Mountains in the spring of 1876.

John F. Casey   1872-88   Born in Caldwell County, Missouri in 1850, barber. (5'8") In June 1877, Casey remembered riding into Musquiz Canyon to relieve a party trapped by Indians. In the early morning darkness, Company H charged through the narrow canyon in "columns of fours." Casey's horse stumbled in a hole and two other horses fell over him, dislocating Corporal Casey's shoulder. Appointed First Sergeant of Troop H at Fort Davis. In October 1884, he went hunting with public women and lied about it to his company commander. Because of this behavior, Casey was brought before a court martial, found guilty, reduced to private and forfeited ten dollars a month for six months. Part of the patrol that capture Mangus. For that action, he was again promoted to sergeant and escorted Mangus to Florida. He was the son of a slave mother and white father, had been married three times and claimed to have been married a fourth time to the infamous Belle Starr. He had two children, Mary and Frank, with his first wife, Pabla. For a while he worked at his own barber shop in Saint Joseph, Missouri. Casey, who had once been a sharpshooter, began to suffer from deteriorating teeth and eyesight. With his third wife, Emma, he spent the last years of his life under the care of the Military Soldiers' Home in Leavenworth, Kansas.

Joseph Cephas   1882-87   Born in 1859 in Warrington, Virginia, farmer. (5'3") On detached duty at Pinery Station in October 1882.

Discharged at Fort Apache. In 1920, he was a widower living in Baltimore with daughter Maggie.

Webb Chatmoun   1884-89   Born in 1863 in Spartanburg County, South Carolina, laborer. (5'6") Enlisted with his older brother, Calvin, on October 16, 1884 in Cincinnati, Ohio. Filed for pension on April 8, 1899 in Missouri. Died on October 6, 1938 at Bevier, Missouri. Minerva Chatman ordered his headstone.

Henry Chesnutt   1882-87   Born in 1861 in Tennessee, farmer. (5'7") Filed for invalid pension in 1890. His widow, Martha Salina Chesnutt filed for a pension in 1902.

Earnest Cherry   1898-99   Born in 1877 at Quincy, Illinois, laborer. (5'4") Discharged at Fort Sam Houston in February 1899.

Albert Christopher   1880-81   Born in 1853 in Boyle County, Kentucky. Arrived from the recruiting depot in February 1881, died on April 8, 1881 at Fort Davis from obstruction of the bowels.

Ellson Clarke   1868   Born in 1846 in Baltimore, Maryland, boatman. (5'4") On the night of November 18, 1868, he entered the house of Capt S.B. Lauffer at Fort Wallace and stole some jewelry from his kitchen. He set fire to the house to cover his theft. Received a dishonorable discharge, and confinement in the penitentiary for two years.

William Clark   1870-73   In April 1873, he got in a dispute at the government ferry for Fort Gibson. The argument escalated and Clark attacked ferryman Charles Reade with an axe. Reade died from his injuries on April 23. Clark was sentenced to be discharged dishonorably and confined to life in prison.

Andy Clayton   1872-82   Born at Carrollton, Missouri in 1851, laborer. (5'8") Suffered from frostbitten toes during the Red River War. In 1874, he was charged with entering the quarters of Lydia Brown, a laundress, and drawing a large knife threatening, "I'll cut you if you don't undress and let me sleep with you." There was insufficient evidence to find Clayton guilty.

Rufus E. Cobb   1898-99   Born in 1877 in Metropolis, Illinois, laborer. (5'10") Died at Rock Island, Illinois on June 10, 1946.

George Coleman   1905-1912   Born in Louisa, Virginia in 1883, cook. (5'8") Served as company cook. Collapsed due to heat exposure during a parade through New York City in 1909.

Joseph Collins   1876-80   Born in 1850 at Providence, Rhode Island, coachman. (5'6") Promoted to sergeant. Later served in the 25th Infantry.

William Cook   1872-82   Born in 1851 in Kentucky, farmer. (5'8") Promoted to sergeant.

Taylor Crockett   1876-86   Born in 1850 in Carroll County, Georgia, laborer. (5'4") After his enlistment, he worked a porter in Prescott, Arizona, remained single. Entered the soldiers' home in Los Angeles in 1923. Died at the home on February 4, 1930 of a cerebral hemorrhage.

John H. Curtiss   1874-79   Born in 1849 in Philadelphia, Pennsylvania, shoemaker (5'5") Served as a saddler in Company H.   Later served in the Ninth Cavalry. Died on July 21, 1879 at Fort Stanton, New Mexico.

Joseph Damon   1884-85   Born in 1859 at Savannah, Georgia, barber. (5'4") Received a disability discharge on March 7, 1885. Lived in Oregon. His widow Edith M. Damon filed for a pension in 1931.

Clinton Davis   1876-81   Born in 1853 at Indianapolis, Indiana, laborer. (5'2") Served as a blacksmith..

Willis Davis   1876-78   Born in 1855 in Upson County, Georgia, farmer. (5'8") Received a dishonorable discharge on June 1, 1878 at Fort Davis.

Peter Dehony   1882-92   Born in 1861 at Columbia, Kentucky, laborer. (5'6") Settled in Indianapolis, Indiana, married, lived until at least 1930.

Curtis DeGroat   1882-92   Born in 1859 in Orange County, New York, waiter. (5'4") Served in Troop E 1882-87 and Troop H 87-92. Filed for an invalid pension on May 31, 1892 in Nebraska.

Frank K. Dickerson   1898-1903   Born in 1865 at Augusta, Georgia, tailor. (5'6") Enlisted in New York City for the Spanish American War, re-enlisted at the end of the war.

James Dillard   1882-93   Born at Allensville, Kentucky in 1861, farmer. (5'7") Member of the patrol that captured Mangus. Drowned in the Missouri River near Fort Buford, North Dakota on July 1, 1893.

Edward Dosier   1898-1905   Born in 1871 at Edgefield, South Carolina, hostler. (5'6") Discharged at Fort Sam Houston in1899, re-enlisted. Served as a farrier.

George Douglass   1872-75   Born in 1847 in St. Clair County, Missouri, laborer. (5'4") Served as a blacksmith. On May 28, 1875, he discharged his pistol at Fort Davis with intent to injure Andy Clayton. He received a dishonorable discharge on July 22, 1875 and confinement at hard labor for one year.

Preston Douglass   1872-74   Born in 1847 in Missouri, blacksmith. (5'7") Sick in hospital in July 1874, received a disability discharge at Fort Sill in August 1874.

Lucelius Drane   1894-99   Born in 1873 in Sumner County, Tennessee, laborer. (5'7") Promoted to sergeant in 1898. Sick in hospital in January 1899. Received a disability discharge on February 19, 1899 in Huntsville, Alabama. Returned to Tennessee and filed an invalid pension on February 27, 1899. His widow, Golden Drane, filed a pension from Illinois on October 30, 1929.

John Dupree   1877-84   Born in 1856 in Lincoln County, North Carolina, laborer. (5'7") Attacked Samuel Porter with a knife at Fort Davis, cutting his head four times and attempting .to slice him in the side. He claimed that Porter had slapped him and spit on the floor near his bunk. He received a dishonorable discharge on October 14, 1884 and six months at hard labor.

James Ecton   1888- 1914 Born in 1869 in Winchester, Clark County, Kentucky, laborer. (5'5") Served as a saddler, promoted to corporal and sergeant. Retired after 26 years of service on August 13, 1914 at Fort Huachuca.

George E. Edwards   1914-1919 Promoted to sergeant. Commissioned a first lieutenant of volunteers during World War I.

Robert Edwards   1898-1905 Born in 1874 in Barnwell County, South Carolina, laborer. (5'4") Wife Mary, settled in South Carolina as a farmer.

Robert Ellis   1888-92 Born in 1866 at Charleston, South Carolina, laborer. (5'5") Found sleeping on duty at Fort Apache in September 1891, received a sentence of four months hard labor and forfeiture of $10 per month for the same period. On February 1, 1892, while still serving a sentence of hard labor in the guardhouse, Ellis received orders from the sergeant of the guard to light the street lamps on post. He ignored the order, later explaining, "It was a moonlit night and the prisoner did not suppose the Sergeant intended what he said when he directed the lights to be lit." Ellis claimed he had already served seven months in the guardhouse when he had only been sentenced to four months. The court sentenced Ellis to an additional two months of hard labor and forfeiture of $10 per month for that period. He deserted in June 1892.

Andrew Emory   1882-87 Born at Richmond, Kentucky in 1861, plumber. (5'7") Enlisted at Cincinnati, Ohio. Had an opportunity to work with Chaplain Francis Weaver to improve the Fort Davis post library.

William Epps   1887-89 Born in 1866 in Richmond, Virginia, laborer. (5'8") In August 1888, while on duty at the San Carlos reservation, he received orders to assist in digging a sink for the detachment. Instead, he wandered away and did not return until other members of the company had finished the work. Refused to return a shovel borrowed from the Twenty-fourth Infantry. Sentenced to hard labor for four months and forfeiture of $10 per month for the same period. Received a dishonorable discharge on June 6, 1889.

William C. Ewell   1887-98 Born in 1866 in Coles Ferry, Charlotte County, Virginia. (5'8") Served as a blacksmith, received a disability discharge at Huntsville, Alabama. Requested an invalid pension in 1905. Worked as a harness cleaner in Richmond, Virginia. His widow, Willie J. Ewell filed for a pension on June 29, 1920.

Charles Faulkner   1879-1909 Born in 1850 in Garrard County, Kentucky. (5'7") Enlisted at Fort Davis in 1879. Served as Quartermaster Sergeant and First Sergeant of Troop H. In 1879, the post surgeon treated him for a sprain resulting from "falling over a tent pin" in the dark. On February 8, 1898 he was named national standard bearer for the Tenth Cavalry. In 1902 he would write, "I have the honor to request that I be allowed to re-enlist in Troop H 10th U.S. Cavalry as a married man (no children), character 'Excellent' and am serving in the 23rd year of continuous service." Retired on October 30, 1909 at the Presidio of San Francisco.

Jackson Ferrer   1874-79   Served as a trumpeter. Cited for gallantry in action at Fort Sill 1874.

Michael Finnegan   1884-94   Born in 1862 in Madison County, Florida. Enlisted at Fort Davis, sharpshooter. In 1890 became First Sergeant of Troop H. Received a disability discharge in 1894, and filed for invalid pension for chronic rheumatism in May 1895 in Missouri. He did receive a small pension for his disability. Finnegan went to sea, working as an Engineer's Steward on ships crossing the Atlantic. In 1914, when war broke out in Europe, he found himself stranded in England without a job. The American Relief Committee of London provided Finnegan passage to New York. Died June 18, 1918 at Hampton, Virginia.

James Fitzpatrick   1876-82   Born in 1850 in Maury County, Tennessee, laborer. (5'7")

Stephen S. Ford   1877-95   Born in 1855 at Warrenton, Virginia, teamster. (5'8") Served his last enlistment with Troop H, before that with Troop B. Promoted to sergeant. Discharged on March 8, 1895 at Fort Buford, North Dakota. Filed for an invalid pension on July 29, 1895 in Minnesota. Worked as a Pullman porter in St Paul, moved to Washington, DC in 1897. His wife Librada filed for a widow's pension in New York on October 3, 1932.

George W. Forster   1870-91   Born in 1851 at Green River, Kentucky. (5'11") On July 25, 1879 a detachment of twelve men led by Captain Michael L. Courtney of the Twenty-fifth Infantry encountered a raiding party between Sulphur Springs and the Salt Lakes. Forster received a gunshot to the left breast. Had a crowbar fall on his hand while laboring on post and sprained his right knee when a horse fell with him in February 1883. By September 1883, he had developed acute rheumatism, requiring regular treatment. Member of the patrol that captured Mangus. Promoted to sergeant. On February 20, 1891, Forster was shot in a dispute with interpreter John Glass at Fort Apache. He died two days later after twenty-one years of faithful service. Buried at Fort Apache, reinterred at Santa Fe National Cemetery in 1932.

William H. Fowler   1876-81   Born in 1853 in Baltimore, Maryland, laborer. (5'6") Discharged at Pena Colorado in 1881.

Henry P. Gaines   1898-99   Born in 1871 in Nevada, hostler (5'7") Enlisted at Fort Snelling, Minnesota, discharged at Fort Sam Houston.

George A. Garfield   1898-1901   Born in 1869 at Natchez, Mississippi, shoemaker. (5'6") Deserted in Honolulu on April 24, 1901. Filed for a pension on March 3, 1902 in Texas.

George Garnett   1872-81   Born in Hannibal, Missouri in 1847. Born a slave in Missouri, enlisted in Company I, 56th USC Infantry September 20, 1863, at the age of sixteen. Promoted to sergeant and served as company first sergeant. Cited for gallantry in action at Fort Sill 1874. Led detachments on the trail of Apaches through the Carrizo and Guadalupe Mountains. Present at the battle of Rattlesnake Springs on August 6, 1880. Received a disability discharge on April 5, 1881. Worked as a gardener in Des Moines, Iowa. Entered the Leavenworth Soldiers' Home in 1905 with rheumatism.

Daniel Garrett   1890-98   Born in 1865 in West River, Maryland, hostler. (5'6") Served as a blacksmith in Troop H, promoted to corporal. Mortally wounded in Huntsville, Alabama on November 11, 1898 by "Horse" Douglass, a local black citizen in the pay of whites who wanted to drive out the Tenth Cavalry. Died on November 14.

William H. Gaston   1897-1907   Born in 1869 at Huntsville, Alabama, teacher. (5'7") Applied for a pension on March 24, 1926 in Minnesota.

William Grafton   1882-87   Born in 1855 at Kansas City, Missouri, laborer. (5'6")

Scott Graham   1881-83   Born in 1854 in Butts County, Georgia, laborer. (5'8") Deserted on November 14, 1883 while rounding up horses in Texas.

General Lee Grant   1909-19   Promoted to corporal, Commissioned second lieutenant of volunteers during World War I.

William F. Grant   1876-78   Born in 1848 in Farquar County, Virginia, shoemaker. (5'6") Died on September 25, 1878 of a knife wound at Fort Davis.

William Graves   1898-99   Born in Limestone County, Tennessee in 1875, railroad brakeman. (5'7")

Charles Gray   1882-92   Born in 1861 at Baltimore, Maryland, laborer. (5'9") Served as trumpeter and sergeant. On October 1, 1889, he was arrested for starting a fight with a civilian employee at Fort Apache. Reduced to private and ordered to forfeit $10. Later served in the twenty-fourth Infantry.

Charles A. Green   1878-86   Born in 1862 in Queen Anne County, Maryland, laborer. (5'5") Member of the patrol that captured Mangus.

James H. Greenfield   1869-74   Born in 1848 in St Mary's, Maryland, laborer. (5'6") Worked as a messenger for the Surgeon General's office in Washington, DC for over twenty years. Married Agnes Williams on November 7, 1882. Filed for a pension in Washington, DC on August 23, 1917. His widow filed for a pension on June 8, 1922.

Thomas Hames   1898-99   Born in 1870 at Russell Cove, Kentucky. (5'5") Served as a saddler. Filed for a pension in Kentucky on November 4, 1901. Died on February 1, 1924. His widow, Hattie Hames filed for a pension in Kentucky on July 3, 1924.

John Wesley Hamil   1894-1902   Born in 1872 at Milner, Georgia, porter. (5'9") Promoted to corporal.

William Hamlin   1868-70   Born in 1844 in Davison, Tennessee, boatman. (5'6") Veteran of the 17th USC Infantry. Caught sleeping while on stable guard at Camp Supply. Deserted in November 1870. Received a dishonorable discharge and three years confinement on November 12, 1870 at Fort Sill.

Washington Hardaway   1879-80   Born in 1855 at Richmond, Virginia, sailor. (5'4") While on herd duty near Eagle Springs, felt his horse stumble and then raise itself up violently. He hurled forward onto the pommel causing a hernia on his left side. Received a disability discharge on September 21, 1880.

George W. Harding   1873-76   Born in 1851 in Orange County, New York, groom. (5'10") Deserted on June 8, 1874, caught the same day. Received a dishonorable discharge on February 18, 1876 at Fort Davis.

Wesley Hardy   1876-80   Born in 1855 in Georgia, farmer. (5'9") Died at the battle of Rattlesnake Springs on August 6, 1880. Scott Lovelace of Company I later reported that Hardy had been carrying a message when the Apaches captured him, "I saw his saddle stripped of all leather. The Indians tied him to a tree and burned him alive.

Augustus Hargrove   1898-99   Born in 1877 in Dallas County, Alabama, moulder. (5'7") Filed a pension in Pennsylvania on April 29, 1929. Worked as a porter in Pittsburg.

Benjamin Harris   1888-91   Born in 1865 in Greene County, Alabama, laborer. (5'8") Pled guilty to public drunkenness, stating, "I feel deeply the disgrace to which the excesses of drink has brought me … the result of unfortunate weakness to which the flesh is heir…"

Hyder Harris   1881-86   Born in 1861 in Caroline County, Virginia, laborer. (5'3") Promoted to corporal. Discharged February 24, 1886 at Bonita Canyon. Filed for a pension on July 5, 1901 in Maryland. His widow, Emma J. Harris, filed for a pension on June 3, 1904.

Lewis Harris   1875-1901   Born in Harford, Maryland in 1847, farmer.. (5'5") Veteran of the 30th USC Infantry. Enlisted at Baltimore, Maryland in April 1875. Served as a farrier, also served in the Twenty-fourth Infantry.

James H. Harris   1869-74   Born in Washington, DC in 1849, laborer. Entered the soldiers home in Danville, Illinois in 1925 with lumbago and arteriosclerosis. Died on January 2, 1926 of heart disease.

Mack Harris   1871-1901   Born in 1850 in Floyd County, Kentucky. Joined Troop H in 1897 with previous service in the Ninth and Tenth Cavalries. Served as a corporal and company cook. Retired on August 31, 1901 at Fort Clark. Filed for a pension in Alabama on April 2, 1919. Lived with his wife, Annie, a private family cook and his father-in-law, a church janitor. Died Huntsville, Alabama on October 21, 1936.

William Harris   1898-1901   Born in 1875 in Nashville, Tennessee, laborer. (5'7") Engaged in an operation to reinforce and resupply Cuban insurgents in 1898.

Edward Hartsfield   1876-91, 1898-1901   Born in 1854 at Raleigh, North Carolina, hostler. (5'6") Also in Troop G, corporal. Filed for a pension on March 11, 1893 in Georgia. Re-enlisted in Troop H in 1898 for the Spanish American War as a sergeant.

Jesse Hawkins   1873-74   Born in 1852 in Louisiana, farmer. (5'6") Deserted on June 8, 1874, caught same day. Received a dishonorable discharge on December 7, 1874.

William Hawkins   1882-87   Born in 1859 in Shelby County, Kentucky, tailor. (5'6") Suffered sunstroke during target practice at Fort Davis in the summer of 1883. Member of the patrol that captured Mangus.

Girard Hayden   1876-79   Born in 1853 in Wayne County, Kentucky, laborer. (5'7") Received a disability discharge on March 1, 1879. Filed for a pension on July 26, 1880 in Kentucky. His widow Lizzie Hayden filed for a pension on December 30, 1929 in Tennessee.

Lewis Hayes   1869-70   Born in Arkansas in 1848, laborer. (5'5") Shot William Shaw in the hand when Shaw was goading him with a horned toad. Then deserted in October 1870. Received a dishonorable discharge on November 20, 1870 and confinement for eighteen months in military prison.

Frank C. Henry   1897-1900   Born in 1872 in Charlottesville, Virginia, laborer. (5'7") Served as a trumpeter. Engaged in an operation to reinforce and resupply Cuban insurgents in 1898.

John Henry   1869-70   Born in 1848 at Chesterville, South Carolina, laborer. (5'8") Stole a carbine at Camp Supply in 1870. Deserted in October 1870. He received a dishonorable discharge on November 4, 1870 at Fort Sill and was sentenced to two years in military prison. Settled in Cross County, Arkansas, married, and worked as a farmer.

James A. Hill 1883-90 Born in Saline County, Iowa in 1861, bartender. (5'6") Served his first enlistment in the 25th Infantry. Enlisted in Troop H at Davenport, Iowa in August 1888. On April 26, 1889 he fell in inspection ranks and had to be carried to the hospital. On May 23, he again suffered vertigo on the porch of the barracks. Received a disability discharge on January 27, 1890 at Fort Apache.

Samuel Hoe 1873-74 Born in 1852 at Harrisburg, Kentucky, waiter. (5'3") Sick in hospital in September 1874. Received a disability discharge on October 11, 1874 at Fort Sill. Filed for a pension on March 26, 1890. Worked as a janitor in Topeka, Kansas. Married Fannie, a laundress. Died at age 90 on April 13, 1942 in Topeka, Kansas.

John Holliday 1876-81 Born in 1851 at Jonesburg, Georgia, waiter. (5'7")

John Hood 1895-1906 Born in 1873 in Greenville, Lawrence County, South Carolina, laborer. (5'7") Later served in the Ninth Cavalry. Received a disability discharge in 1906.

James Hopper 1894-1909 Born in 1870 at Mount Airy, North Carolina, teamster. (5'6") Promoted to sergeant. Died on September 23, 1909 at Fort Ethan Allen.

George Horton 1873-89 Born in 1850 in Winnsboro, South Carolina, laborer. (5'8") Served as a saddler. Had been previously enlisted in the 9th Cavalry.

Frank Howard 1876-81 Born in 1855 at Greenville, Georgia, laborer. (5'7") Discharged on December 3, 1881 at Pena Colorado.

Moses W. Hull 1888-98 Born in 1864 at Augusta, Georgia, cook. (5'5") Worked as a porter at a wholesale house in Atlanta, wife Anna, son Henry. Filed for a pension on August 5, 1930 in Georgia. Buried on January 18, 1940 in Marietta, Georgia.

Dickson Hunter 1876-77 Born in 1852 in Virginia, laborer. (5'8") Received a dishonorable discharge on November 2, 1877 at Fort

Davis and confinement in the military prison at Fort Leavenworth for six years.

George Hunter 1872-73 Born in 1852 in Platte County, Missouri, laborer. (5'11") Accidentally shot at Fort Sill, died in the hospital on August 7, 1873.

John Hurt 1898-99 Born in 1871 at Hopkinsville, Kentucky, laborer. (5'5")

Abraham Jackson 1897-1907 Born in 1874 at Washington, DC, coachman. (5'7")

Edward Jackson 1869-74 Born in 1848 at Richmond, Virginia, laborer. (5'7") Promoted to sergeant.

Henry Jackson 1898-99 Born in 1869 in Virginia, hostler. (5'4") In charge of the mules during the attack on the heights at Santiago, Cuba.

Isaac Jackson 1873-88 Born in Montgomery County, Kentucky in 1853. (5'7") Promoted to sergeant.

Isaac B. Jackson 1882-83 Born in 1857 in Jamaica, cook. (5'8") Deserted on September 17, 1883.

James Jackson 1876-81 Born in Virginia in 1842. In May 1881, he suffered a contusion to his right arm after being "kicked by a government mule, in the line of duty".

Charles A. Jefferson 1873-92 Born in 1852 at Baltimore, Maryland, laborer. (5'4") Lost his new break open Smith & Wesson revolver. Also served with the Ninth Cavalry.

Andrew J. Jennings 1898-1901 Born in 1874 in Wilson County, Tennessee, farmer. (5'7") Promoted to sergeant. Received a dishonorable discharge on February 6, 1901 at Manila.

Isaac Jernigan 1898-99 Born in 1875 at Paris, Tennessee, laborer. (5'6")

Dorsey Johnson   1879-83   Born in 1860 at Wolf Creek, Kentucky, laborer at a tobacco factory. (5'9") Sprained his left knee jumping off a horse. Died of an inflammation of the lung leading to pneumonia at Fort Davis on March 5, 1883. Buried in San Antonio. His mother, Georgia, barely managed to survive working as a chambermaid on Mississippi steamboats. She filed for and received a pension as a military dependent on August 21, 1885.

Edward Johnson   1873-78   Born in 1852 in Queen Anne County, Maryland, laborer. (5'5") Transferred to the Ninth Cavalry in 1878, died of a stab wound in the abdomen on December 24, 1878 at Fort Bayard, New Mexico.

George H. Jones 1868-69   Born in 1847 at Somerset, Maryland, laborer. (5'5") Deserted on August 6, 1869, surrendered, deserted again on September 27, 1869.

John Kimber   1883-88   Born in 1860 in Robinson County, Tennessee, laborer. (5'8")

Ellison Kitt   1895-98   Born in 1868 in Knoxville, Tennessee, laborer. (5'7") Served as a saddler. Filed for a pension on May 8, 1922. His widow, Narsisia, filed for a pension on July 21, 1930 in California.

Andrew J. Knight 1902-05   Born in Meriweather County, Georgia in 1872, student. (5'6") Placed second in the broad jump in a 1903 army track meet.

Robert Lang   1891-1908   Born in 1855 at Cynthiana, Kentucky, brick mason. (5'6") Promoted to sergeant. Sick in hospital in the Philippines, Camp John Hay in June 1907. Retired at the Presidio at San Francisco on September 29, 1908.

Frank Lanier   1879-84 Born in 1856 in Nashville, Tennessee, laborer. (5'7") Applied for a pension in March 1927 in Illinois.

Charles Lann   1898-1900, 1916-1919   Born in 1875 at Cincinnati, Ohio, laborer. (5'5") Received a disability discharge on March 29, 1900. Filed for a pension in May 1906. Married Elsie and lived in Chicago. Served in the 370th Infantry during World War I. Died on September 22, 1936.

Douglas E. Lee   1876-81   Born in 1853 at Columbus, Ohio, laborer. (5'7")

John E. Lewis   1896-1914   Born in 1879 in Roanoke, Virginia, laborer. (5'9") Promoted to corporal. Served as a correspondent to the Richmond *Planet* during the Spanish-American War.

John Lisby   1876-77   Born in 1850 in Maryland, farmer. (5'5") Died of typhoid on June 29, 1877 at Fort Davis, buried in San Antonio.

Lewis H. Logan   1869-72   Applied for a pension in March 1927 in Oklahoma.

James T. Lott   1905-14   Born in 1884 in Runge, Texas, laborer. (5'5") Member of the machine gun platoon. Deserted in Aug 1914.

Leywood Loving   1897-1900   Born in 1875 in Nelson County, Virginia, laborer. (5'10") Engaged in an operation to reinforce and resupply Cuban insurgents in 1898.

John Mack   1869-74   Born in 1846 at Boonville, Missouri, laborer. (5'7")

Louis Mack   1869-74   Born in 1848 in Clay County, Missouri, laborer. (5'6") Promoted to sergeant. Cited for gallantry in action at Fort Sill 1874. He was wounded during the action.

Thomas Maddox   1878-84   Born in 1852 in Grayson County, Texas. (5'9") Promoted to sergeant. Died of an abscess of the liver at Fort Stockton on June 1, 1884.

John Madison 1884-92 Born in 1850 in Shelbyville, Kentucky, porter. (5'7") Served his second enlistment in the 24th Infantry.

Nat Mahue 1876-81 Born in 1855 in Edmundson County, Kentucky, laborer. (5'5") On guard duty at the Corsence Ranch in March 1881.

Colonel E. Miller 1879-92 Born in 1843 in Elkton, Maryland, sailor. (5'9") Member of the patrol that captured Mangus. Promoted to sergeant. Kicked by a mule in the left knee and suffered a back injury playing baseball. Struck in eye by a cartridge shell at target practice in May 1887. Married Rose, had three daughters and two sons. Settled in Feliciana, Louisiana.

Robert Miller 1873-78 Born in 1852 in Boone County, Missouri, laborer. (5'7") Alexander Brown forced Miller to desert with him. Testifying against Brown at the court martial, he declared, "He asked me was I going to desert with him. I told him, yes. He had a pistol in his hand cocked." Miller was acquitted and returned to duty.

Doc Mocfield 1876-91 Born in 1855 in Charlotte, North Carolina, cook. (5'7") Sick in hospital in July 1881. Drunk and disorderly at Fort Apache in 1887. He was confined at hard labor for four months and forfeited $10 per month during that period.

John M. Morgan 1876-77 Born in 1855 at Newberry, South Carolina, laborer. (5'8") Sick in the hospital with a cough and returned in the evening, shot in the abdomen. Died November 27, 1877 at Fort Davis, buried in San Antonio.

Simon Motlow 1898-99 Born in 1872 at Lynchburg, Tennessee, farrier. (5'9") Applied for a pension in May 1925. Died on February 26, 1930, buried in Chattanooga, Tennessee.

John Muchs 1873-83 Born in 1851 in Anderson County, Kentucky, laborer. (5'9") Died of pneumonia (inflammation of lungs) at Fort Davis on August 2, 1883, buried San Antonio. His mother Nancy Meaux filed for a dependent pension on September 11, 1885 in Kentucky.

William Neal   1898-99   Born in 1874 in Hamilton County, Ohio, laborer. (5'7")

George H. Newman   1877-1907   Born in Middleburg, Virginia in 1859. (5'7") Promoted to corporal. Member of the patrol that captured Mangus. Also served in the Ninth Cavalry and 24th Infantry.

Edward W. Nimms 1898-99   Born in 1877 at Cincinnati, Ohio. (5'8")

William Nugent   1884   Born in 1854 at Evergreen, Louisiana, blacksmith. (5'4") Deserted on August 7, 1884.

George Osborne 1907- 1935   Born 1886 in Lexington, Kentucky, farmer. (5'6") He married Vesta Monroe from New York in 1919. Worked as a civilian at Fort Ethan Allen, Vermont until 1954. Died of cardiac arrest on April 26, 1983 at the age of 98

Alfred Owings   1869-71   Born in 1848 in Orange County, Virginia, sailor. (5'7") Died on February 18, 1871 of wounds caused by the accidental discharge of a carbine at Fort Sill, buried there.

William Patterson   1898-99   Born in 1874 in Henry County, Tennessee, laborer. (5'7")

Joseph F. Pendleton   1876-81 Born in 1853 in Louden County, Virginia, laborer. (5'7")

Charles Perkins   1882-83   Born in 1853 in Bradon County, Kentucky, cook. (5'6") Received a dishonorable discharge on July 18, 1883 at Fort Davis.

Albert Pierson   1876-81   Born in 1855 at Monroe, Georgia, laborer. (5'5") Settled in Thomaston, Georgia with his wife Estelle and granddaughter Julia working at the cotton mill.

William Plumo   1887-99   Born in 1867 at Charleston, South Carolina, sailor. (5'8") Served as a trumpeter.

Samuel Porter   1877-87   Born in 1855 at Nashville, Tennessee, laborer. (5'6") Drew his pistol in a quarrel over a card game. He was confined at hard labor for one year and forfeited $10 per month for same period. In another incident at Fort Davis, John Dupree attacked Porter, cutting his head because Porter had slapped him and spit on the floor near his bunk.

William K. Porter   1896-1902   Born in 1875 at Paris, Kentucky, laborer. (5'7") Promoted to corporal. Engaged in an operation to reinforce and resupply Cuban insurgents in 1898. Settled in Pittsburg, Pennsylvania, married Loucelia, worked as a city patrolman. Died July 16, 1962 in Pittsburg.

Frank Posey   1881-88   Born in 1860 in Montgomery County, Maryland, laborer. (5'7") Enlisted on November 21, 1881 at Baltimore, Maryland. He deserted on February 27, 1885. Enlisted in the Twenty-fifth Infantry under the name, Frank Brocko. On May 15, 1888, he was recognized as a deserter and arrested at Fort Snelling, Minnesota. Received a dishonorable discharge, forfeiture of all pay and confinement for two years at hard labor. Based on his service record, under both names, the court granted clemency and reduced the sentence to one year hard labor. They later commuted the sentence following a request from the governor of Minnesota stating that Posey had a wife in Saint Paul "who is sickly and is in destitute circumstances.

Benjamin Raysor   1887-92   Born in 1864 in Orangeburg, South Carolina, farmer. (5'7") Promoted to corporal. With his wife Mary, he applied for a pension on March 3, 1927, died on March 15, 1935 in Branchville, South Carolina.

Charles Reed   1883-97   Born in 1850 at Louisville, Kentucky, laborer. (5'6") Enlisted at San Antonio, Texas. Received a disability discharge on July 10, 1897 at Fort Assiniboine.

Ollie Rodgers   1898-99   Born in 1880 at Paducah, Kentucky, laborer. (5'9") Filed for a pension on May 11, 1912 in Kentucky.

Alexander Ross 1879-84 Born in 1853 in Madison County, Kentucky, laborer. (5'10") Served as a farrier.

Joseph Rousey 1876-91 Born in 1855 at Stanton, Virginia, laborer. (5'3") Member of the patrol that captured Mangus. Settled in Winslow, Arizona.

Clifford A. Sandridge 1897-1919 Born in 1874 in South Charleston, Ohio, farmer. (5'8") Promoted to sergeant, served as First Sergeant. Commissioned Captain in October 1917, assigned to the Meuse-Argonne sector during World War I.

Romeo Satterthwaite 1879-84 Born in 1853 in Pitt County, North Carolina, laborer. (5'10") Promoted to corporal. On an extended border patrol, he became very ill with bronchitis and rheumatism from exposure. He remained in the hospital for several months and continued to have problems with coughing and shortness of breath afterward.

Greenfer Shanklin 1873-76 Born in 1851 in London, Ontario, Canada, farmer. (6') Enlisted at Detroit. On December 8, 1875, he was drunk on duty and resisted arrest. Received a dishonorable discharge and confinement in the penitentiary for two years.

Robert L. Shell 1898-99 Born in 1870 at Greenville, Tennessee, laborer. (5'9")

Daniel Simpson 1876-81 Born in 1855 at Rockfield, Kentucky, laborer. (5'8") Promoted to sergeant.

Charles H. Smart 1899-1919 Born in 1880 at Dinnsboro, South Carolina. Promoted to sergeant. Shot in the toes of his left foot during the fighting at Naco, Arizona in 1914.

Ephriam Smith 1868-73 Born in 1843 at Richmond, Virginia, blacksmith. (5'7") Convicted of insubordination at Camp Supply in 1870, confined at hard labor for one month.

George H. Smith 1898-99 Born in 1874 at Louisville, Kentucky, laborer. (5'7")

Albert H. Squiers 1888-1901 Born in 1867 at Prescott, Wisconsin, hostler. (5'5") Enlisted in Troop H with prior service in the 24th Infantry. Filed for a pension in December 1925 from Canada.

Augustus Sparks 1885-95   Born in 1856 at Nicholasville, Kentucky, laborer. (5'7") Promoted to corporal and sergeant. Member of the patrol that captured Mangus. Also served in the 24th Infantry.

George Stacks 1875-80 Born in 1852 in Todd County, Kentucky, laborer. (5'5") Discharged in the field on September 12, 1880. Later worked as a cooper in Louisville, Kentucky.

Thomas Straws 1879-84 Born in 1852 at Frankfort, Kentucky, laborer. (5'3") His father farmed near Frankfort.

William M. Strayhorn 1896-99 Born in 1874 at Durham, North Carolina, laborer. (5'5") Served as a farrier. Filed for a pension on January 28, 1927. Died on March 19, 1931 in Hampton, Virginia. His widow Minnie filed for a pension on April 13, 1931.

James Stewart 1881-83 Born in 1855 in Montgomery County, Virginia, laborer. (5'9") Killed by the accidental discharge of a carbine, died on April 9, 1883.

Charles I. Taylor 1892-1900 Born in 1871 at Smithland, Kentucky, laborer. (5'6") Also served in the 25th Infantry. Promoted to corporal. Later worked as a manager in the Negro baseball leagues. He and his wife Cora took in old soldiers including Fitz Lee, who they cared for during his last months.

Charles Terry 1884-1911 Born in 1865 in Warsaw, Indiana, barber. (5'6") A member of the patrol that captured Mangus. Served as a trumpeter, went to Cuba with the Tenth Cavalry band. Wounded in the side during the assault on Santiago, he returned to his home in Indianapolis to recuperate. Received a disability discharge in 1911 for pulmonary emphysema. Married Grace Clayton in 1919, died November 9, 1951 in Delaware.

William Thacker 1888-1912   Born in 1861 in Lawrenceburg, Kentucky. (5'6") Served in Troops F and G as well as H. Retired as First Sergeant of Troop H on December 11, 1912.

Alfred Jack Thomas 1903-1919   Born in 1881 at Pittsburgh, Pennsylvania, musician. (5'7") Transferred to the Tenth Cavalry Band in 1906 and rose to the position of chief musician and band director. Also served as the band director for the 368th Infantry in World War I. Settled in Baltimore, Maryland and opened the Aeolian Conservatory of Music, also composed music. He died of a stroke on April 19, 1962.

Perry Thomas   1876-81   Born in 1848 in Queen Anne County, Maryland, waiter. (5'7")

Sharp Thomas   1882-88   Born in 1855 in Nottoway County, Virginia, farmer. (5'3")   Began his second enlistment with the 25th Infantry, received a disability discharge on March 13, 1888.

Alfred Thompson   1893-99   Born in 1870 in Wilson County, Tennessee, porter. (5'8") Served his second enlistment in the 24th Infantry. Discharged with rheumatism on January /24, 1899.

Benjamin Timbers   1896-1903   Born in 1874 in Maryland, tailor. (5'7")

George H. Turner   1896-99   Born in 1869 at Brooklyn, New York, porter. (5'5")

Simon Turner   1873-83 Born in 1852 in Lancaster County, Virginia, laborer. (5'6") Promoted to corporal.

Robert Valentine   1869-71   Born in 1848 at Alexandria, Virginia, laborer. (5'9") On May 10, 1870, he threatened Edward Jackson with a loaded carbine and managed to shoot a horse by accident. Sentenced to forfeit $10 per month until the $177 price of the horse was paid and six months at hard labor. Deserted on July 21, 1871.

Henry Walker   1881-85   Born in 1859 at Shelbyville, Kentucky, laborer. (5'6") Enlisted in Louisville as a blacksmith. Had a horse fall on his foot and a few months later he smashed his finger with an axe. In November 1884, a piece of iron struck him in the right eye. He lost all use of that eye and received a disability discharge on March 18, 1885.

Benjamin F. Wallace 1880-85 Born in 1853 in Chester County, Pennsylvania, shoemaker. (5'5") Served as a saddler.

George H. Washington 1892-1913 Born in 1866 at Washington, DC, teamster. (5'3") Served as a farrier and sergeant. Re-enlisted in the Ninth Cavalry. Received a disability discharge in 1913. Filed for a pension in May 1913. Died on December 28, 1945, buried at Arlington.

Jacob Watkins 1876-86 Born in 1847 in Burton County, Kentucky, farmer. (5'6") Veteran of the 110th USC Infantry. Enlisted at Bowling Green. Promoted to corporal.

Asa Weaver 1876-81 Born in 1853 at Randolph, Indiana, laborer. (5'7") Led the patrol that located Victorio's Apaches on August 3, 1880. Fought a fifteen mile running battle to report to Captain Carpenter. Received an immediate promotion to sergeant for his cool handling of the situation and his complete report on Victorio's position. In the fall of 1880, he developed trouble with his vision while patrolling the Rio Grande. Weaver's sight became so poor that he was relieved of guard duty "and had to be led about at night." Returned to Rush County, Indiana, married Catherine in 1882, had a daughter Myral A. Weaver, worked as a day laborer.

William J. Webb 1873-80 Born on June 1, 1854 in Toronto, Canada, farmer. (5'7") Settled in Trenton, New Jersey in 1869 and enlisted in December 1873. Promoted to corporal and sergeant. On July 25, he was wounded when his detachment encountered a raiding party between Sulphur Springs and the Salt Lakes. He was shot in the right thigh and never fully recovered. Later he was tying down a horse when it reared up and kicked him in the head just above the right eye. He also strained the cords in his right hand on saber drill. Received a disability discharge on January 18, 1880 at Fort Davis.

Sampson West 1881-93 Born in 1857 at Lexington, Kentucky, gardener. (5'10") Enlisted at Chicago. Discharged on February 24, 1893 at Fort Buford.

John E.N. Westfall 1898-99 Born in 1868 at Newton, New Jersey, trainman. (5'6") Served as troop clerk in 1898-1899.

John W. White 1898-99 Born in 1870 at Washington, DC, cook. (5'7") Served as company cook.

Robert White 1867-82   Born in 1846 in Marion County, Virginia, servant. (5'9") Previously served in the 38th and 24th Infantry. Promoted to sergeant. With the company at the battle of Rattlesnake Springs on August 6, 1880.

George Wilkerson 1879-84 Born in 1858 in Scott County, Kentucky, laborer. (5'6")

Henry Williams 1882-84 Born in 1861 in Montgomery County, Tennessee, laborer. (5'8") Received a disability discharge on March 1, 1884 at Fort Davis.

James Williams 1880-97 Born in 1848 at Wooster, Ohio, teamster. (5'7")

William Willis 1876-81 Born in 1855 in Alabama, farmer. (5'4")

Anderson Wilson 1869-72 Born in 1848 in Platte County, Missouri, farmer. (5'5") Caught sleeping on duty at Camp Supply in 1870. He received confinement at hard labor for six months and was fined $10 per month for the same period. On June 21, 1872, while on patrol, he managed to shoot himself in the hand when picking up his gun. When sent to the end of the column, he deserted. He received a dishonorable discharge and confinement for two years in the military prison at the Texas State Penitentiary in Huntsville.

Thomas Wilson 1877-91 Born in 1853 in Maryland, farmer. (5'7") Enlisted at Baltimore.

Walter Wilson   1894-99 Born in 1868 in Ablemarle County, Virginia, laborer. (5'8")

Henry Winship 1868-69 Born in 1847 in Clinton, Georgia, laborer. (5'11") Deserted on August 6, 1869.

Edward Wiser   1881-86 Born in 1860 in Montgomery County, Missouri, laborer. (5'6") Enlisted at St Louis, discharged at Fort Apache. Served as a blacksmith.

Clark Wright   1884-89   Born in 1864 at Glenville, New York, laborer. (5'3") Enlisted at Indianapolis, discharged at Fort Apache.

Warren W. Wright   1869-78   Born in 1846 at Maysville, Kentucky, barber. (5'6") Enlisted at Philadelphia. Promoted to sergeant. Deserted and received dishonorable discharge on September 2, 1878.

Joseph Wooden   1877-1904   Born in 1860 in Glasgow, Kentucky. (5'7") Promoted to corporal and sergeant. Retired on March 5, 1904.

Charles Woods   1869-74 Born in 1841 in Louisville, Kentucky, farmer. (5'8")

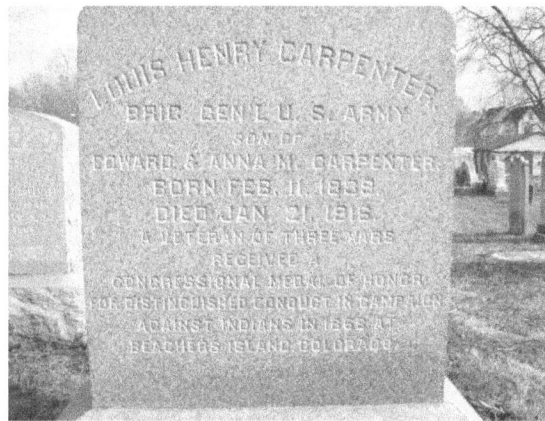

INDEX